Basic Accounting

J Randall Stott and
Mike Truman

Revised by

Andy Lymer and
Nishat Azmat

For UK order enquiries: please contact Bookpoint Ltd,
130 Milton Park, Abingdon, Oxon OX14 4SB.
Telephone: +44 (0) 1235 827720. Fax: +44 (0) 1235 400454.
Lines are open 09.00–17.00, Monday to Saturday, with a 24-hour
message answering service. Details about our titles and how to
order are available at www.teachyourself.com

Long renowned as the authoritative source for self-guided
learning – with more than 50 million copies sold worldwide –
the Teach Yourself series includes over 500 titles in the fields of
languages, crafts, hobbies, business, computing and education.

British Library Cataloguing in Publication Data: a catalogue record
for this title is available from the British Library.

This edition published 2010.

Previously published as Teach Yourself Basic Accounting.

The Teach Yourself name is a registered trade mark of
Hodder Headline.

Copyright © 1985, 1997, 2003, 2010 J Randall Stott,
Mike Truman, Andy Lymer and Nishat Azmat

Typeset by MPS Limited, A Macmillan Company.

Printed in Great Britain for Hodder Education, an Hachette UK
Company, 338 Euston Road, London NW1 3BH, by CPI Cox &
Wyman, Reading, Berkshire RG1 8EX.

The publisher has used its best endeavours to ensure that the URLs
for external websites referred to in this book are correct and active
at the time of going to press. However, the publisher and the
author have no responsibility for the websites and can make no
guarantee that a site will remain live or that the content will remain
relevant, decent or appropriate.

Hachette UK's policy is to use papers that are natural, renewable
and recyclable products and made from wood grown in sustainable
forests. The logging and manufacturing processes are expected to
conform to the environmental regulations of the country of origin.

Impression number	10 9 8 7 6 5 4 3 2 1
Year	2014 2013 2012 2011 2010

Front cover: © Darby Sawchuk / Alamy.

Back cover: © Jakub Semeniuk/iStockphoto.com, © Royalty-Free/
Corbis, © agencyby/iStockphoto.com, © Andy Cook/iStockphoto.com,
© Christopher Ewing/iStockphoto.com, © zebicho – Fotolia.com,
© Geoffrey Holman/iStockphoto.com, © Photodisc/Getty Images,
© James C. Pruitt/iStockphoto.com, © Mohamed Saber – Fotolia.com

Contents

Meet the authors

In this book we will introduce you to the core concepts and practices of accounting. It is assumed you have little or no prior knowledge of this topic and so we will teach you the basics to enable you to produce simple accounts for a business or for a club or other not-for-profit group. Correspondingly, we also provide you with the basic knowledge needed to be able to read and understand the accounts of other businesses so you can see how they operate and can therefore assess whether or not you want to do business with them, become a shareholder, loan them money or undertake the myriad of other business-to-business transactions you may be considering and need financial data to help you make decisions on.

Clearly this book cannot teach you everything you might need to know in a few hundred pages. However, if you work through all that is written here and undertake the exercises given and the examples shown, on completion of the book you will be well on your way to becoming an accounting literate person with the ability both to run your own business more successfully, and engage more efficiently with other businesses.

You may be studying for a qualification in accounting – perhaps struggling with the reading materials proposed by your course leaders. We hope the very readable and engaging approach we have adopted will help you understand what you need to be successful in your courses. We have specifically included revision exercises throughout to make this book help you in preparing for examinations of your newly gained knowledge.

If you find you need more detail on specific aspects of accounting principles or practice that are not covered in adequate depth here then we recommend you look at a companion book in this series (also part written by one of the authors of this title) – *Get to Grips with Book Keeping*. Conversely, if the accounting principles outlined seem to be more complicated than your simple business requires – or you need more information on how to use the accounting figures for tax purposes or other regulatory 'red tape' requirements – then check out *Teach Yourself Small Business Accounting*, again part written by the same team with a different slant on the materials that may be more suited to your needs.

This book was originally written by J. Randall Stott, and updated in recent editions by Mike Truman; we gratefully acknowledge the foundation work done by these prior authors on whose significant efforts we have tried to build further as part of our development and updating of this title for this edition.

We hope you enjoy the book and gain in business confidence as a direct result.

Andy Lymer and Nishat Azmat, September 2009

Only got a minute?

In simple terms, 'accounting' is the process of collecting, measuring, recording and communicating financial information about a business to the variety of users who want access to this information. It is a cyclical process that operates alongside the actual activities of a business to provide a summary view of what is going on and to help understand as much of its financial aspects as possible. It involves collecting all the transaction data that have financial implications for the business's assets, liabilities, or the owner's money invested in the business, then providing a measurement of each of these transactions in some way, so that they have financial value/costs associated with them. The next step is recording them in a way that others will understand (often, although not necessarily always using a system called double entry bookkeeping) and then using these records to summarize the transactions into various forms for different users, both internal and external to the business.

Once you have an overview of each of these stages and how accounting practices operate, you can understand how to do your own accounting, and also how to understand and interpret the accounts prepared by other businesses you may want to deal with.

5 Only got five minutes?

The collection, measurement, recording and communication of financial information about a business to users to help them understand what has been going on within a business is the foundation of the accounting process – at least from a financial perspective. These principles and practices of accounting are operated by the majority of businesses around the world.

The core principles of how to account for a variety of businesses – from the very smallest sole traders, through organizations like charities and other 'not-for profit' organizations, into partnerships and up to companies – are easier to understand than you might think. While all of these groups have key differences in their setup and structure that make them different legally, they all follow a recognizable (albeit increasingly complicated) accounting process that is based on the same rules and processes introduced in this book.

The place to begin is with the collection and recording part of this process of accounting – where does accounting information come from and what are the basic principles of double entry bookkeeping that are used to record these in a systematic, structured, reliable and checkable way?

Once the basic practices of accounting have been understood, the newcomer can move on to more detailed accounting principles – accounting for stock, dealing with checking bank statements, managing credit accounts and operating cashbooks before looking at tax issues likely to be necessary to account for (VAT and PAYE).

Of course, putting information into an accounting 'system' is of little use to anyone unless you also create effective ways of getting the data out again – ideally in summarized forms that help pinpoint the specific aspects of the business's activities you may be interested in understanding at the time. The accounting system enables us to extract statements summarizing profits or losses incurred by the business over time. It also describes what the financial position of the business is (at any point in time; what assets it owns or can use in its business activities and how these assets are funded). In the latter case this is usually a mixture of the owner's investment plus other loans made to the business in cash or in goods and services as they trade on credit. This is all neatly summarized for us in what is called the Balance Sheet.

These basic accounting principles and practices learned initially, illustrated for simple, sole trader, businesses can be readily extended for different types of business. This includes amendments for partnerships (e.g. extra issues then arise to do with how profits are split between the different owners using capital and current accounts, applying rules agreed in partnerships agreements, etc), for not-for-profit entities (e.g. use of receipts and payments accounts and income and expenditure statements instead of profit and loss accounts) and also the different rules that apply once a limited liability business is established and wider regulation for how its accounting has to be undertaken become applicable (e.g. tracking its share capital, detailing specific statutory reporting requirements and so on).

Wider aspects of accounting practices that might also form part of a wider understanding of accounting include:

▶ social and environmental reporting – as examples of non-financial reporting that many businesses are increasingly providing (particularly larger businesses) tell users important things about how the business is operating
▶ computerized accounting – basic accounting principles continue to apply once computers are used for at least part of the practice of accounting, hence it is critical you understand these principles even when using a computer to do much of the actual data management work

▶ the impact of accounting standards on accounting practices – particularly the role of IFRS for larger businesses and the FRSSE for smaller businesses.

Knowing all of this will clearly not make you a fully functioning accountant – you need to pass lots of hard exams for that! – but it will, we hope, give you the confidence to be able to 'talk accounting' both as it applies to your own business, and also in equipping you to analyse what others are showing with their accounts as you make decisions about the variety of interactions you have with other businesses (e.g. as potential or actual shareholders, debtors or creditors and possible providers of loans, etc.).

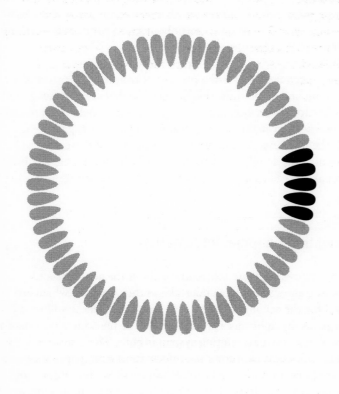

10 Only got ten minutes?

A thorough understanding of accounting – both its principles and its practices – is a vital tool in running a successful business. To be able to prepare basic accounts for a small business – but also to be able to understand the accounts produced by other businesses you may be aiming to do business with as a trading partner, invest in, loan money to and so on – can only increase your confidence and your chances of business success.

What is accounting? It is the process of collecting, measuring, recording and communicating financial information about a business to those who need that information to help them manage the business, or make other decisions about their engagement with the business. As you work through each of these stages of accounting in detail your knowledge will build but here is a brief summary of each stage so you can see what is coming:

Collecting financial information

All businesses generate information about the activities and various transactions they undertake as they operate, which are useful for the accounting function to use. A business will try to capture as much of this as it can to provide the source information on which its own accounting system is built. These sources primarily include capturing the information from paper trails, e.g. invoices, receipts, credit notes, wages slips, time sheets, job cards and so on, although increasingly these days it also means capturing electronic transactions from sources such as internet sales. While accounting systems are usually full of internally generated data, at times they may also include external sources of information.

Measuring financial information

Once the source documents have been captured financial values need to be attached to them so they can be included in an accounting system. Often this is straightforward and the transaction will have a financial value already associated with it, such as a price or quantity. However, this isn't always the case so rules are needed for how to give values to transactions when that situation occurs.

Only the information captured into a system can be reliably measured, of course. This is why the accounting system is never a complete reflection of the entire business in 'all of its glory'. Some aspects of a business can't easily (or at least with acceptable reliability and agreement) have financial values associated with them, such as the value of good employees, or their brand. However, enough of the normal activities of the business can usually be measured to make the accounting system provide a fairly comprehensive view of what has been going on. However, don't forget, as you get into the details that the accounting system is only an abstracted view of some aspects of a business. It is never a complete window into the whole business. This is why, for example, a company's share price never perfectly reflects a fixed percentage of what the accounts say a business's net assets total up to ('net' here meaning after all its debts have been paid off). A business is unlikely to be worth exactly the total of all its assets minus its debts – the bits the accounting system tells us about. It is often worth more as it is a functioning and viable business, although sadly it can be worth less if it isn't viable! These measurement and valuation aspects of accounting systems are explored further throughout the book.

Recording financial information

The key focus of basic accounting is the process used to record financial information. For a majority of businesses this is a system called double entry bookkeeping – a 500-year-old way of recording business transactions that still works perfectly well today for many businesses. Double entry bookkeeping, however, is not the only system that can be used to record such activity. So while this method is the most popular way of doing accounting, it is also worth looking at simpler accounting systems that may work perfectly well if your business recording needs are relatively simple.

Grasping how double entry bookkeeping works is the core of understanding accounting and, with practice, you will be able to understand these rules easily. It is not rocket science even though many people often find it a bit confusing initially!

Communicating financial information

Once as much of the business's activities have been as reliably captured, measured and recorded as we can you need to be able to access and use this information 'database' for it to actually be helpful in managing. Therefore, following on from the focus on learning bookkeeping, the next key task is exploring the various ways in which financial information can be summarized by, and extracted from, the accounting system (e.g. profit and loss accounts, and balance sheets illustrating respectively the performance of a business over time and its financial position at a specific point in time). It can also be often important to understand the wider reporting that may be undertaken by a business that extends these basic summarizing statements for other uses, e.g. cash flow statements, extracting information for managers to use, creating control accounts and how the external reporting environment impacts on the ways in which reporting to external users occurs.

Focus

For most people interested in using accounting the main focus is likely to be profit seeking (i.e. 'for-profit') businesses, so using sole traders as the core business model is probably the best place to start in seeking to learn about basic accounting. However, once you have understood these core rules you will see that very similar principles can be applied to other for-profit business structures (partnerships and limited liability companies) and even to not-for-profit 'businesses' like clubs, societies and charities. You can learn the key differences for doing and understanding accounting for each of these different businesses (e.g. use of different owners' equity structures such as capital and current accounts for partnerships, share capital, share premium and other reserves for limited liability companies) and different forms of reporting 'surpluses' if your business model is not about making 'profits' in the normal sense.

Other topics

An exploration of a few wider topics in accounting is also worthwhile perhaps including how businesses (particularly larger ones) also report on some non-financial matters to try to broaden the information that they communicate about their business activities that 'normal' accounting does not reflect fully.

Understanding accounting standards and how these rules for how businesses must do their accounting apply in practice for many businesses is also particularly important to a full grasp of basic accounting.

Using a computer as a part of an accounting system changes how you operate but it is important to note that it does not alter the need to be able to understand and correctly apply the general processes of accounting, particularly the collection, measuring and communicating areas of these elements as these continue to operate in largely the same way even if a computer can help you with the routine recording of data.

You can learn to be comfortable with 'talking accounting' by working through a book like this. This alone will not make you a professional accountant. Although anyone can actually call themselves an accountant in the UK, most qualify by taking exams to demonstrate the knowledge they have, as well as undertaking practical experience, so you'd ideally need to obtain professional qualifications if you want to get a job applying the knowledge you have gained. However, even if you have no intention to become an accountant, understanding the basic principles of accounting will enable you to better understand what your accountant tells you, to appreciate what they are doing for you, or to do more (perhaps even all) of your own accounting for your business (if you are brave enough!). You can also learn about other businesses by reading their accounts. This should help you greatly in all your interactions with other businesses – either as one of their owners/shareholders, as a provider of loans or as part of trading relationships (e.g. as a debtor or creditor).

1

Introduction

In this chapter you will learn:
- **the purpose of accounting**
- **the meaning of assets and liabilities**
- **what a balance sheet looks like**

There is nothing magical about bookkeeping and accounting. The *keeping of accounts* simply involves the routine recording of business transactions, the exchange and the payment for both goods and services being identified by money values.

The accounting device and method explained in this book is called 'double entry bookkeeping'. Sometimes referred to as the *arithmetic of commerce*, it is a form of simple arithmetic – that is all.

Of course, accountancy is more than double entry bookkeeping, just as mathematics is more than simple arithmetic. This book is intended for those with no previous knowledge of the subject, and covers the topics needed for a basic accounting course. It also introduces some of the subjects required at higher levels, such as social accounting and cash-flow statements.

However, preparing for examinations is not the only reason for reading this book. You may work in the accounts department of a business and want to understand how the work that you do fits in with the overall accounting system of the business. You may be a manager who wants to understand the accountancy implications of the decisions that you make.

Alternatively you may own a business and want to know how to keep your business records. If you have a very small business and operate as a sole trader you may find the system outlined in another of our titles, *Teach Yourself Small Business Accounting*, is easier to follow. However, as soon as you have a lot of sales or purchases on credit, or a lot of plant and machinery, or several bank accounts then you will need to understand the underlying principles of accountancy outlined in this book. If you need more detail of accounting principles and the practical operations of bookkeeping than are provided here, do explore *Get to Grips with Book Keeping*.

The purpose of accounting

We live in a world of credit. Without some form of accounting record, indecision and confusion would result. Not only is it businesslike to keep accounts as a check on suppliers and credit customers, but the financial statements made up from these accounting records keep management informed, from time to time, of the progress of their businesses.

The financial statements are scrutinized and verified when:

▶ *a business is sold or a new partner is admitted to an existing firm*
▶ *a bank loan or a substantial overdraft is needed by the business*
▶ *HM Revenue and Customs enquires into a business's tax liability.*

Business transactions involving all aspects of services, production, trade and distribution are varied and voluminous in every town and village. The bulk of these transactions are on credit, with payment for the goods bought or for the services used often being delayed for a few days or a few weeks. It is important that there is electronic or written evidence of the original terms and money values agreed upon between producer and consumer, vendor and credit customer, or the professional service-provider and the client.

In double entry bookkeeping, the money values of all transactions with suppliers of goods on credit (the **creditors**, also known as payables) and the sales to credit customers (the **debtors**, also known as receivables) are recorded. The details of all money received and money paid, whether by cheque or in cash, are also recorded in a manner which, with practice, becomes routine and easy to comprehend.

Some everyday terms have a different meaning when used in the accounting sense. A few of these terms are now introduced.

▶ **Assets** *refer to the property and possessions of a business, which may be owned by an individual (a sole trader or one-person business), a firm or partnership, or perhaps a limited company. In this book, bookkeeping is explained as it applies to the financial accounting side of a sole trader's business. Nevertheless, basic principles of double entry, throughout the business world, remain the same.*
  ▷ Fixed assets *(also known as non-current assets) are those assets which are retained for the benefit and permanent use of the business, such as premises, machinery and plant, vehicles, fixtures and fittings. These assets are not for re-sale in the trading sense.*
  ▷ Current or circulating assets *change their form in the course of trading, common examples being stock of goods for sale, the trade debts of customers, money in the bank and office cash.*
▶ **Liabilities** *are the debts and obligations of the business, the external and trade liabilities falling into two main categories:*
  ▷ fixed *or* non-current liabilities *such as a mortgage or a long-term loan, not due within 12 months.*
  ▷ current liabilities *comprising outstanding accounts owing to trade and expense creditors, and sometimes a bank overdraft, payable within 12 months.*

In addition to the external debts and obligations of the business, there is generally a large internal debt owing by the business to its proprietor under the heading of **capital**. This is sometimes called the owner's 'equity' in the business.

Balance sheet: an introduction

The first consideration to be given to any new business venture is that of finance. A trading business needs substantial funds or extended credit facilities from the outset. A place has to be found for the storage and safety of merchandise, and thought needs to be given to deliveries (involving transport), communications (post, telephone and email), and the recording of cash and credit dealings (the book work and accounts). In addition, at least some small reserve of finance is required to maintain oneself during the initial period of creating or developing the business.

Before we start on the routine book work of a typical small trader, we shall take a glimpse at one of the targets and main end-products of this short course in elementary accounting. This is a financial statement called the **balance sheet**.

> **Insight**
> A balance sheet is a snapshot in time looking always at past events, i.e. transactions that have already taken place. An accounting period of 12 months is generally used for this type of financial reporting. Users of this type of information include management, owners, employees, lenders, etc.

Suppose your great-aunt Sarah recently died and left you £5000. You decide to be 'your own boss' and start a business from your home address.

With effect from 1 June, this £5000 is allocated to your new business venture to become the sole asset and property of the business in your name. The business, regarded as separate from you personally, acknowledges its debt to you as owner and proprietor in this opening balance sheet, thus:

Balance Sheet as at 1 June

Assets employed	£	You may have other sources of income and property of your own, but that is your private affair, quite distinct from this new trading venture now being financed by your investment of £5000.
Cash in hand	5000	
Financed by	£	
Proprietary capital (owner's equity)	5000	

The routine business dealings of a small retail trader are called transactions. Generally, they involve the purchase and sale of merchandise for cash and/or credit (where payment is delayed), the settlement of trade and expense accounts for goods bought and for services used, and occasionally the purchase of a 'fixed asset' for permanent use in the business.

During the first week of June a number of transactions take place, and in this particular instance a separate balance sheet has been drawn up simply to illustrate how this financial statement is *affected in two ways by each transaction*. Normally, though, the listing and grouping of assets and liabilities on a balance sheet would be made in greater detail at the end of the trading period, perhaps every six months or only once a year.

2 June You pay £1300 for storage cupboards and some strong shelving.

NB The four stages of these elementary balance sheets are explained by simple arithmetic. The routine debit/credit procedure starts in Chapter 4.

Balance Sheet as at 2 June

Assets employed		£	Property has been acquired for permanent use by the business. This purchase is shown as a 'fixed asset'. Cash is decreased by the sum paid out.
Fixed assets			
Fittings/fixtures		1300	
Current assets			
Cash in hand		3700	
		5000	The payment for the fixed asset does not affect the holding of the proprietor (his capital) nor current liabilities, as the business at this stage has no outside debts.
Financed by		£	
Proprietary capital		5000	
Current liabilities		5000	

3 June You buy goods on credit priced at £2000 from Wholesalers Ltd, arranging to pay for the goods bought later in the month.

Balance Sheet as at 3 June

Assets employed			Goods to the value of £2000 have been taken into stock at cost price, increasing current assets to £5700. These goods have been bought on credit (no money has been paid). So cash remains at £3700. There is still no change in the owner's capital account. However, an outside liability has been incurred of £2000. The individual names of trade creditors do not appear on the balance sheet.
Fixed assets		£	
Fittings/fixtures		1300	
Current assets	£		
Stock	2000		
Cash in hand	3700	5700	
		7000	
Financed by		£	
Proprietary capital		5000	
Current liabilities			
Trade creditors		2000	
		7000	

4 June You sell goods priced at £600 to a cash customer, and further goods priced at £400 to a credit customer (to be paid at the end of the month).

Say you make a profit of 20% on the selling price. The cost of the goods sold is thus £800.

Balance Sheet as at 4 June

Assets employed			£800 of stock (at cost) is sold for £1000. This leaves stock balance in hand at £1200. The cash position is now £3700 + £600 = £4300. Amount due from credit customer (£400) is shown under 'trade debtors'. The difference between the cost price £800 and the selling price of £1000 is the **trading profit**. This is added to the proprietor's capital as the reward for investment.
Fixed assets		£	
Fittings/fixtures		1300	
Current assets	£		
Stock	1200		
Trade debtors	400		
Cash	4300	5900	
		7200	
Financed by			
	£		
Capital 1 June	5000		
Add profit	200	5200	
Current liabilities			
Trade creditors		2000	
		7200	

5 June You pay £500 off your supplier's account for the goods bought on 3 June, and then withdraw £250 cash for your own private and personal use.

Balance Sheet as at 5 June

Assets employed			
Fixed assets		£	
Fittings/fixtures		1300	
Current assets	£		
Stock	1200		
Trade debtors	400		
Cash	3550	5150	
		6450	
Financed by			
	£		
Capital 1 June	5000		
Add profit	200		
	5200		
Less drawings	250	4950	
Current liabilities			
Trade creditors		1500	
		6450	

The £500 paid off trade creditor's account reduces both the business cash and the amount of the trade liabilities total. The stock figure is not affected by this payment. The sum of £250 withdrawn for the proprietor's own use is called 'drawings'. Business cash is reduced and also the proprietor's holding or net assets as shown by his capital account.

Note how the proprietor's capital account of this small business remains constant until affected by:

▶ *business profits or losses*
▶ *withdrawals by the proprietor (and it would of course be increased by additional private capital paid into the business).*

Some elementary accounting concepts have been touched upon in this short balance-sheet preamble. At each stage there is the emphasis upon total assets equalling total liabilities (including the capital). The accounting equation $A = C + L$ applies all the way through, where:

A *represents the total assets of the business*
C *the proprietary capital and*
L *the external debts and liabilities of the business.*

Insight

Remember that the balance sheet reflects the accounting equation and there are several ways to write the accounting equation, which means there are several ways to present the balance sheet. You must take this into account and be flexible when you see different financial statements.

TESTING YOURSELF

1.1 Capital = £50 000

Liabilities = £20 000

How much are the assets?

1.2 Fixed Assets = £65 000

Current Assets = £20 000

Long-term Liabilities = £30 000

Current Liabilities = £15 000

What is the capital?

2

Source documents

In this chapter you will learn:
- *the source documents from which accounting records are prepared*
- *the differences between sales, purchases and banking records*

In order to complete the accounting records, a bookkeeper needs to have source documents to work from which show what the financial effect of the transactions is. The way in which the documents are used is explored in more detail in Chapter 14, but the documents themselves are introduced here. They come in many different shapes and sizes, but they can be split into three categories.

Sales documents

The key document relating to credit sales made by the business is the invoice. When a business delivers goods or provides services and allows the purchaser time to pay (in other words, makes a credit sale) it will issue an invoice. This sets out the goods or services provided, gives the name and address of the purchaser as well as that of the supplier, states the amount that is due (separately identifying any VAT) and normally gives a date or time limit for payment.

But what if a customer complains that, although you have billed him for 1000 widgets, you only sent 900? The customer may issue a 'debit note', which formally sets out the shortfall and the amount that the customer thinks should be offset against the bill. Whether he does or does not issue a debit note, your response (if you think the complaint is justified) is to issue a credit note. This looks like an invoice, but states that the amount owing to you by the customer is to be reduced by the amount shown.

You may well have encountered credit notes when shopping, if you take goods back. Here you have already paid for the goods, so the credit note is money which can be taken off your NEXT purchase. In the business world, most of the time the credit note is raised BEFORE the bill is paid, so the amount is taken off the amount paid for the CURRENT purchase. The principle, however, is exactly the same.

On a practical level, the original invoices and credit notes are of course sent to the customer. You may be working from a copy, or in some cases simply from a computer listing of transactions within the accounting system. You should also note that the seller may send out monthly statements of the amount owing, or reminders for overdue invoices; these should not be confused with invoices, even though they may look similar.

A business which sells mainly for cash may still prepare invoices, or receipts which are very similar to invoices. For most retail businesses, however, the main document which accountants will work from is the till roll. This shows the total amount taken in for goods and services, whether by cash, cheque or card.

Purchase documents

In a very small business which buys everything for cash, the only record of a purchase will be the till receipt for payment. If the payment was made by cash, there may also be a petty cash slip – see Chapter 13. However, larger businesses need a more formal system to ensure that purchases are authorized by the right person and that payment for the goods is only made when they have been received and checked.

Assume that the Maiden Megastore is ordering 5000 CDs from Acme Artistes:

1 *A purchase order is prepared by Maiden Megastore – a form that is sent to Acme Artistes setting out the 5000 CDs to be supplied and the price to be paid (already agreed between the companies). This has to be authorized at an appropriate level of management within Maiden.*
2 *Acme sends the 5000 CDs, and at the same time encloses a 'Dispatch note' giving the details of the shipment. This will be checked against the order and approved by Maiden.*
3 *Acme then sends an invoice to Maiden for the agreed price of the 5000 CDs.*
4 *Maiden checks the invoice against the purchase order, and may also check it against the dispatch note. The aim is to ensure that the goods were genuinely ordered, and have been received. A failure in the system at this point would mean that an unscrupulous trader could simply issue invoices for non-existent goods or services to businesses taken from a trade telephone directory and get paid! If there have been any problems with the shipment, debit and credit notes may be issued.*
5 *Finally (and possibly after Acme has had to issue a statement of the account, see above) Maiden sends Acme a cheque for the amount owed, often using a payment slip that can be torn off the invoice. It is not normal in business for Acme to then issue a receipt for the cash received if it was in response to an invoice. Acme will enter the payment against the amount owed by Maiden, leaving nothing outstanding, and will pay the cheque into the bank.*

Note that the purchase order and dispatch note are not records from which accounting transactions should be entered, since they do not create or satisfy financial obligations. They are useful when calculating accruals (these are operating expenses owing at the end of a period) for final accounts, but the main reason for including them here is to alert you to their existence so that you do not confuse them with other documents.

Banking documents

The final category of documents used to create the accounting records is banking documents. Most of these will be familiar from your own personal banking.

The bank statements are crucial to the preparation of accounting records. They may be the only record of some transactions, such as payments made direct into the bank account of a supplier by a customer through the banking system (a 'BACS' transfer). In other cases they will summarize the information that is available elsewhere.

Payments made by cheque will generate two different records. The customer should have a completed cheque stub (often called a counterfoil) in the cheque book. The supplier should have a completed paying-in slip stub/counterfoil in the paying-in book, showing the payment of the cheque into the bank.

Sometimes cheques are dishonoured by the customer's bank. They will then be returned to the supplier who paid them in, marked 'refer to drawer'. If you look back at the example of Maiden and Acme you will see that when Maiden first received the cheque from Acme this was entered into their accounting records, with the result that Maiden is now shown as owing nothing. If the cheque is returned dishonoured it has proved to be worthless, and another entry reversing the payment will have to be made in Acme's books so that Maiden is once again shown as still owing the money due for the 5000 CDs. A similar adjustment will have to be made in the books of Maiden to reflect the fact that they still owe Acme the price of the CDs.

TESTING YOURSELF

2.1 Explain the following terms:

a *credit note for sales returned*
b *credit note for purchases returned*
c *accrued expenses*
d *an invoice*
e *a dishonoured cheque.*

3

The ledger system

In this chapter you will learn:
- *the meaning of ledger account and cash account*
- *how to enter basic transactions*

The ledger system of double entry bookkeeping involves the use of a number of account-ruled books (known as a set of books) for the purpose of recording accurate information, in money values, of the day-to-day trading operations of a business. From these permanent records, periodical statements are prepared to show the trading profit or loss made by the business and its assets and liabilities, at any given date.

In the past, these records would quite literally have been kept in bound ledger books. However, even before the widespread introduction of computers, mechanized systems based upon mechanical accounting machines were used by many larger companies. In smaller organizations, looseleaf systems with multipart forms and carbon paper reduced the number of times that bookkeepers had to write out the same data.

Now any business with a full-time bookkeeper is likely to have computerized its accounting. However, computerization can only speed up the arithmetic of accounting, it cannot replace an understanding of the concepts. Underlying all modern computer accounting programs is the same double entry system described in this book.

Ledger accounts

The record of trading transactions is kept on the folios or pages of these account books, called **ledgers**. The ledger folios have special rulings to suit the needs of the business. The bank statement style (page 85) lends itself to modern accounting, but for the time being double entry will be explained by the older traditional method. This is the ordinary ruling for the older style of ledger account:

Name of Account

Debit or left-hand side			Credit or right-hand side		
Date	Value COMING IN	£	Date	Value GOING OUT	£

In practice, separate ledgers are kept for the different classes of accounts (customers, suppliers, business property, trading revenue and expense, etc.). Batches or groups of similar accounts are kept together, and ledgers are indexed so that information with regard to any particular account may quickly be obtained.

Insight

It is important to check the accuracy of entries made in ledger accounts at regular intervals as failing to debit a customer's account for goods bought on credit, for instance, could result in not sending out a reminder for payment.

The cash account

The cash account is first explained as *part of the ledger system*. Some students may already be familiar with the simple recording of receipts and payments of money.

Assume that you start in business (a small retail shop) on 1 June with £2000 *cash as your starting capital*. At present we shall only deal in currency notes and coin: cheque and bank transactions will be explained later.

During the month of June all your business dealings are for cash. The details of your trading transactions are listed below, and then entered or 'posted' in your **cash account** for the month of June as the first stage of a bookkeeping exercise involving the *receipt and payment of money only*. Note that at this stage there are no credit transactions.

June		£
2	Goods bought for re-sale	750
5	Money received for goods sold	458
12	Paid for classified advertising	73
18	Cash takings (i.e. cash sales)	562
20	Bought further goods for cash	850
24	Cash sales	1105
30	Paid salary to temporary assistant	600
	Withdrew cash for private use	
	(this is called **drawings**)	500

The cash account is now made up for the month of June, all money received being debited, and all money paid being credited. Note the one-word description of all the credit entries, and the complete absence of the words 'cash' and 'paid' on this side, because all these entries refer to cash payments.

Cash Account

Dr. (debit)   (credit) Cr.

Date	Money COMING IN	£	Date	Money PAID OUT	£
June			June		
1	Balance in hand	2000	2	Purchases	750
5	Cash sales	458	12	Advertising	73
18	Cash sales	562	20	Purchases	850
24	Cash sales	1105	30	Salary	600
			30	Drawings	500

Explanatory notes

1. *The cash in hand, a debit balance at 1 June, is your starting capital in this particular instance, being the sole asset and property possessed by the business on this date. In later exercises it will be seen that business property in money is normally* only part of the proprietor's capital.

2. *Cash takings or cash sales refer to the normal and continuous 'across the counter' sales of the small retailer, in contrast to any* **credit sales** *where the possession of the goods passes from vendor to buyer at the time of the sale, with the settlement and payment being deferred until later. In these early exercises we are concerned only with cash purchases and cash sales.*

3. *The narrow column on the left of each cash column is used for cross-referencing to the appropriate and corresponding account of the double entry (explained in Chapter 5).*

4. *Goods and merchandise bought by the business to be re-sold are called* **purchases**, *distinguishing this cost from the other many and varied expenses of trading, and also to show a clear distinction between this basic trading expense and the cost of property and assets such as machinery and fixtures.*

5. *The one-word description, as far as possible, of each entry, is important, as this serves to identify the opposite and corresponding double entry, normally in another ledger.*

6. *The posting of all amounts* coming into *an account on the* **debit side**, *and all amounts* paid or going out *of an account on the* **credit side**, *avoids the necessity for continual addition and subtraction until the balancing-up stage, generally at the end of the month.*

7. *The shape formed by the line which runs across the top of the account and the line running down between the credit and the debit side is the reason why these are popularly known as 'T accounts'.*

Insight

The main rule for all cash accounts, is that you debit cash coming in and credit cash paid out.

TESTING YOURSELF

3.1 You are required to post up the transactions listed below in a cash account for the first week of March.

March		£
1	Balance of cash in hand	400.00
3	Cash takings	264.40
4	Paid for further goods	148.00
	Paid advertising account	105.00
5	Cash sales	460.20
	Bought further supplies	166.00
6	Wages paid for temporary assistance	250.00

3.2 Give three good reasons for the permanent recording of business transactions.

3.3 What is the difference between a cash sale and a credit sale?

3.4 What is the difference between a cash purchase and a credit purchase?

3.5 Define trade receivables (debtors) and trade payables (creditors).

4

Balancing the cash book

In this chapter you will learn:
- *how to balance the cash book*
- *how to carry down the balance*

In bookkeeping the term 'balancing' simply means adding up both the debit and the credit sides of an account, and deducting the smaller side (of less total value) from the larger side. The difference between the two sides is called the **balance** of the account.

Later it will be seen that the cash account is kept in a special ledger called the cash book, which, in practice, would probably be balanced weekly, and certainly at the end of every month.

The cash account on page 18 is reproduced here and balanced up in the ordinary way. Study the illustration and then read carefully the following instructions on balancing. The payments shown are after any discount allowed – this is explained in Chapter 14.

Cash Account (1)

Dr. (debit)   (credit) Cr.

Date June	Money COMING IN		£	Date June	Money PAID OUT		£
1	Balance in hand	2	2000	2	Purchases	5	750
5	Cash sales	15	458	12	Advertising	8	73
18	Cash sales	15	562	20	Purchases	5	850
24	Cash sales	15	1105	30	Salary	9	600
				30	Drawings	4	500
				30	Balance c/d		1352
			4125				4125
July							
1	Balance b/d		1352				

Note that there are fewer items on the debit of this cash account, and spaces have been left blank to allow for neatness and to ensure that the *corresponding totals are on the same horizontal level*.

The **folio** or page numbers relating to the opposite and corresponding double entry have been inserted in the narrow folio column in front of each amount, prior to the posting of the double entry in the separate ledger accounts as shown in the next chapter. For clarity, the folio or page number is shown in brackets after the name of each ledger account in these early stages.

The balance of cash in hand at 30 June (£1352.00) is the difference between the debit total of £4125.00 and the total of the payments (£2773.00). This balancing figure of £1352.00 is inserted as an additional item on the *credit side above the total*. The two totals now agree and are ruled off in the manner shown. The total of the payments (£2773.00) before balancing may be noted in pencil, but *is not inked-in* as a permanent feature.

The difference or balance on an account should never be left suspended in mid-air. In the case of the cash account, the balance will be entered as the last item on the credit side above the total, and then *brought down below the debit total on the opposite side*.

The two totals are ruled off neatly on the same horizontal level, the lower line of the total being double ruled.

The abbreviations **c/d** and **b/d** signify 'carried down' and 'brought down', and where balances or totals are sometimes carried forward from one folio to another, the abbreviations **c/f** and **b/f** denote 'carried forward' and 'brought forward'.

Insight

The physical cash balance must always be a *debit balance* as money can only be paid out of an available fund or balance in hand. You must never have a credit cash balance.

KEY POINTS

▶ First make your additions in pencil and ink-in afterwards. In balancing, remember to deduct the total of the payments from the total of the receipts, only inserting the difference (which is the balance) on the credit side above the total on the right-hand side of the cash account.

▶ Bring down the same amount shown on the credit side above as a debit balance now below the total on the left-hand side of the account. This debit balance is simply the excess receipts over payments.

▶ Cash account totals must be on the same horizontal level, and the lower line should be double-ruled to indicate a total.

▶ With a little concentration, these simple cash book exercises will quickly teach you how to record elementary cash transactions, leading on to the routine of ledger posting.

▶ In Testing yourself 4.2, although payments are mentioned to a supplier and by two credit customers, we are only concerned at this stage with how the money received and the money paid affects your cash position. The earlier part of this kind of transaction (the credit purchase and the credit sales) is explained later in the text.

TESTING YOURSELF

4.1 What is the purpose of the two columns on the left and the right of the ledger, debit always being on the left and credit always being on the right?

4.2 Post up the transactions shown in a cash account for the month of September, rule off the account, and bring down the new balance as at 1 October.

Sept.		£
1	Cash balance	2000.00
2	Bought goods for cash	356.00
4	Receipts from customers	421.00
8	Paid rent for month	300.00
10	Cash takings	105.00
	Paid advertising account	164.00
12	Cash sales to date	382.80
	Stationery supplies bought	126.00
18	Cash takings	284.80
20	Further purchases for re-sale	193.60
22	Paid carriage on purchases	22.50
	Paid part-time assistant	560.00
28	Withdrew for private expenses	300.00

4.3 At the close of business on 30 June you have £3000 in cash, your only asset and commencing cash balance for the month of July.

Your daily cash sales for July total £1685. Two old customers pay their outstanding accounts (Marcia Moore £234 and Lesley Lane £356) on 5 July. You pay your own supplier Robert Reese & Son

£525 on 20 July for deliveries during June. Other payments made in July were:

July		£
3	Rent	400.00
11	Petrol/oil	125.00
15	Stationery	83.60
28	Salaries	480.00

On 25 July you withdrew £500 from the business to go on a theatre trip. From this information, you are asked to post up a cash account for the month of July bringing down your cash balance on 1 August.

5

Double entry theory and practice

In this chapter you will learn:
- *the principles of double entry bookkeeping*
- *ledger accounts for capital, purchases, etc.*
- *how to record money taken out of the business*

The cash account shown in Chapter 4 is two-sided like all ledger accounts, but this is not the meaning of double entry bookkeeping. Each entry in the cash account, whether receipt or payment, has only one aspect.

The cash account itself is a useful device for recording money received and money paid. Balancing it up will reveal and check the physical cash in hand, but the cash book figures alone cannot be used as a true guide to finding out the trading profit or loss of a business.

To do this, we now need to identify and record the other side of each cash account entry as it is posted to another ledger account.

First principles of double entry

In all business transactions there are *two separate and distinct aspects* to be recorded within the same accounting period.

Two separate accounts are involved in each transaction, one account recording the debit (receiving) aspect, the other recording the credit (giving or paying) aspect.

> **The account** receiving value or benefit is debited.
> **The account** giving, paying or relinquishing benefit is credited.

Sometimes it is easier to decide or identify the account (or person) giving up something of value. This will determine the *credit part* of the two aspects and at the same time serve to identify the *related debit*.

Name of first account

Dr.							Cr.
Date	**Debit what comes into the account** (value received)		£	Date	**Credit what goes out** (value paid or relinquished)		£

Name of second account

Dr.							Cr.
Date	**Debit what comes into the account** (value received)		£	Date	**Credit what goes out** (value paid or relinquished)		£

PRACTICAL BOOKKEEPING

In the previous chapter, only one aspect of double entry was recorded with regard to each item in the cash account. The opposite aspect of each entry is shown on the related ledger accounts.

The reason for the one-word description of most of the postings in the cash account will now be understood, as these are the *names and headings of the new ledger accounts*.

In practice, the postings from the cash account to the related ledger accounts would be made in date order, but for ease in explanation, all debit items will be dealt with first, and then the credit items.

Capital and net worth

The first entry in your cash account is the commencing cash in hand, a balance of £2000 at 1 June. You have given or loaned this business £2000, which from this date becomes the sole asset and property of the business. This is also your own *personal contribution* to the business, and in consequence will be credited to your **capital account** to show that you now have a financial stake in this business of £2000 at 1 June.

Thus the posting of this first item of £2000 to the credit of capital account initiates the double entry to the opposite side of another ledger account.

Capital Account (2)

Dr.							Cr.
				June			£
				1	Cash	CB1	2000

Cash sales or daily takings

The remaining three entries on the debit side of the cash account are all for cash sales or daily cash takings. These individual daily sales are posted to the **credit side of sales account** for the month of June, and in due course this grouped information will be used to work out the trading profit for the month. In practice, the postings would be made day by day.

Sales Account (15)

Dr.							Cr.
				June			£
				5	Cash	CB1	458
				18	Cash	CB1	562
				24	Cash	CB1	1105

Now we turn to the *credit* side of the cash account where the nature of the entries is a little more varied. All items on this side refer to payments for purchases, trading expenses incurred, whilst 'drawings' indicates that some money has been taken out by the proprietor for personal or private use.

Goods bought for re-sale

Goods may be bought for stock or making into products for sale. In any event, a clear distinction must be made between goods bought as the stock in trade of a business and the permanent property (assets) and everyday trading expenses of the business. Goods bought for re-sale are termed 'purchases' and thus the payments made on 2 June and 20 June are posted to the *debit* side of the purchases account.

Purchases Account (5)

Dr.							Cr.
June			£				
2	Cash	CB1	750				
20	Cash	CB1	850				

Trading expenses

The various trading expenses for the month of June (in this instance only advertising and salaries) are debited to *separate accounts*, completing their double entry.

Advertising Account (8)

Dr.   Cr.

June			£				
12	Cash	CB1	73				

Salaries Account (9)

Dr.   Cr.

June			£				
30	Cash	CB1	600				

Money withdrawn for private use

The remaining credit item of £500 for drawings is not a business expense. It is a withdrawal by the proprietor of part of the money invested in the business, although in the case of profitable businesses the owner would expect this to be an advance of future profits. Nevertheless, since part of the business assets (cash) is reduced, there is also a reduction of the proprietor's holding (capital account). Sums withdrawn will temporarily be debited to the drawings account, and at the end of the trading period be *transferred* to the debit of the capital account.

Drawings Account (4)

Dr.   Cr.

June			£				
30	Cash	CB1	500				

In this small exercise and introduction to double entry the word 'cash' has been used throughout in the body of each account, simply indicating that the original posting or entry has come from the cash account (or **cash book** as it will shortly be called). At this elementary stage, the cash book is our main and only **book of original entry,** but as we look at more complex situations, it will be seen that the initiation stage is not always from the cash book.

▶ *Consider now the use and purpose of the folio columns, those narrow columns in front of the main cash columns. What is the accounting term used for entering or making the dual entry from one account to another?*

▶ *How would you decide which account should be debited and which account should be credited?*

▶ *Is it possible to finish with the cash balance brought down underneath the total on the credit side of the cash columns?*

Basic double entry bookkeeping is at the heart of accountancy across the world, right up to the largest multi-national companies. Experienced accountants discussing the effect of transactions will frequently resort to sketching out handwritten T accounts to confirm their impact. Double entry bookkeeping is the international language of accountancy, and it takes practice to become fluent in it.

Insight

You will need a lot of practice in order to become comfortable with double entry bookkeeping. Don't worry if, as you go through the following chapters, you initially find them hard work. The principles of double entry bookkeeping are not familiar to most people, but students generally find that after a while it suddenly 'clicks' and they can handle the basic transactions easily and quickly. Don't give up if your first few weeks of learning seem to be very difficult!

KEY POINTS

▶ Financial information is being grouped at this stage, in total form, for the use of the accountant and management. By this means, the task of the accounts department is made easier in the preparation of various financial statements.

▶ Do not cramp the writing-up of your ledger accounts in an effort to economize on paper. Leave a space of several lines between accounts. Again, remember to leave a space between each ledger account heading and the details in the account.

▶ Do not scratch out or try to rub out any wrong figures. Use 'white out' liquid very sparingly, and be careful to let it dry before writing over it. In practice, though, it is often best simply to rule through incorrect amounts, inserting the correct figure above the error.

▶ As far as possible, reduce the description of the money value received or paid to one word, as this will be the heading of the opposite or contra account as a rule.

▶ Use the £ sign only at the top of the cash columns, and do not repeat the name of the month once inserted at the top of the debit or credit columns within the same account.

▶ Number the various ledger accounts in Testing yourself 5.2 and give some thought to the cross-referencing procedure between cash book and ledger.

TESTING YOURSELF

5.1 Study the cash account of Mark Clark now shown, and then answer the questions underneath.

  a *What is meant by 'Balance b/f'?*
  b *Explain each item in date order.*
  c *Distinguish between cash sales and the sales to credit customers.*
  d *Why is information about purchases shown separately from other expenses?*
  e *Again, why are the payments for salaries and 'drawings for self' not combined?*
  f *Balance up this cash account, and post up the double entry to the corresponding ledger accounts.*

Cash Account

Dr.			£	June			Cr.
June							
1	Balance b/f		1750.00	2	Purchases		1200.00
9	Cash sales		398.00	5	Adverts		156.00
14	Cash sales		664.50	9	Garage a/c		128.50
22	Cash sales		523.50	18	Self		300.00
28	Cash sales		844.00	25	Salaries		1122.50
				30	Rent for June		350.00

5.2 Make up the cash account of Mohammed Khan's part-time business from the cash transactions, shown below, for the month of January.

Jan.		£
1	Cash in hand	250.00
2	Bought goods to be sold	66.60
	Paid for stamps	8.20
5	Cash takings	54.48
12	Paid salaries	64.00
	Withdrew for 'self'	30.00
14	Sundry expenses	14.15
18	Cash takings	84.94
	Paid insurance	12.50
22	Drawings	30.00
28	Paid salaries	64.00
30	Telephoned Thomas Hudson for urgent delivery of stock	
	Consignment received the next day, invoiced at	£44.50

After balancing up Mr Khan's cash account for January, post each double entry to the related account in the ledger.

5.3 What are the bookkeeping entries for the following:

a *a decrease in a liability*
b *an increase in revenue*
c *an increase in an asset?*

6

The trial balance

In this chapter you will learn:
- *how to take the account balances to a trial balance*
- *what to do if the trial balance doesn't!*

The next stage, after the completion of the postings, is the extraction from the books of all balances on a statement called the **trial balance**. It is *not* an account, just a list of all the debit and credit balances.

The financial information, classified and grouped in the various ledger accounts in the last chapter, is now totalled on each account, and the debit and credit *balances* listed on the trial balance, including the *final balance of the cash account*.

Trial Balance 30 June

		Debit £	Credit £
Cash balance	CB1	1352	
Capital	L2		2000
Sales	15		2125
Purchases	5	1600	
Advertising	8	73	
Salaries	9	600	
Drawings	4	500	
		4125	4125

There is no particular order for the grouping of the account balances, but to avoid omission it is suggested that the final cash balance should be extracted first, and the remainder of the ledger balances may then be listed in either page or book sequence.

There is no complication about double entry here, at this stage it has been completed. Debit balances are merely listed on the debit of the trial balance, and credit balances on the credit.

Purpose of the trial balance

It is emphasized that the trial balance is not an account. It is not part of the double entry, but merely a list of debit and credit balances taken from the books, normally at the end of a specific trading period (which may be a month, six months or often one year) for the purpose of:

- ▶ *checking the arithmetical accuracy of the postings, and*
- ▶ *rendering the work of the accountant or bookkeeper easier in the making up of financial summaries, often with a view to the preparation of revenue trading accounts, together with a balance sheet.*

Note that only the balances of the accounts are bought on to the trial balance. For example, the debit balance of £1352 is extracted from the cash account, and *not the two totals* for receipts and payments.

The sales account is totalled to show total sales of £2125 for the month, which in effect is the credit balance taken to the trial balance. Similarly with the purchases, and so on.

If no mistakes have been made in posting the cash book to the various ledger accounts (debit for credit and vice versa), the sum total of the debit balances on the trial balance should equal the sum total of the credit balances.

TRIAL BALANCE TOTALS DO NOT AGREE

If the trial balance totals do not agree, you should try to find the error. It is often useful to calculate the difference between the totals. You may find that this gives a figure that you can find in the original list of balances and which you have either not included in the trial balance, or have not included in your addition of the trial balance figures.

If this is not the case, try halving the difference (if it is 'even') and seeing if a balance of that amount has been included in the wrong side of trial balance, where it will have a double impact in the discrepancy.

Finally, try dividing the difference by nine. If it divides exactly, you may have made a transposition error, for example entering 1985 rather than 1895.

Insight

Errors which will leave the trial balance totals equal are as follows:

▶ *mistakes of the same amounts in both the debit and credit entries*
▶ *omitting the transaction all together*
▶ *the wrong ledger being debited with the correct amount.*

Errors which will cause the trial balance totals to be unequal are as follows:

▶ *arithmetic error when totalling the balances of the ledger*
▶ *correct amount on one side and an incorrect amount on the other, i.e. DR 540 but CR 504 by mistake*
▶ *not including both entries, i.e. only writing entry on one side, and omitting the other*
▶ *including both entries on the same side, i.e. instead of a DR and a CR you have two CR entries.*

TESTING YOURSELF

6.1 Assemble the information shown below in trial balance form, and ascertain the capital of this business at 31 December by deducting the total credit balances from the total debit balances.

	£
Cash in hand 31 Dec.	225
Advertising expenses	65
Purchases for December	4500
Sales for December	5750
Rent and rates paid	360
Stationery bought	40
Salaries paid	880
Drawings for month	400

6.2 Enter up the transactions shown below for a small part-time business in your cash book and post the double entry to the ledger accounts. Balance the accounts where necessary, and prove the arithmetical accuracy of your postings by taking out a trial balance at 31 January.

Jan.		£
1	Cash in hand	300.00
2	Stationery	15.50
	Bought goods	120.00
6	Paid rent for market stall	30.00
8	Cash sales	38.80
11	Classified advertising	10.50
15	Paid wages	60.00
	Bought stamps	5.00
18	Cash sales	56.20
22	Purchases	84.00
24	Petrol	18.30
	Cash sales	42.10
28	Wages	60.00
31	Drawings	50.00
	Cash sales	64.40

6.3 Explain the difference between salaries and drawings. Why is HM Revenue and Customs very much concerned about this distinction when checking the sole trader's year-end accounts?

6.4 What is the main purpose of the trial balance?

6.5 How can you ascertain whether your trial balance has a transposition error? Explain transposition.

Revision exercise 1

In this exercise you start with £400 cash as the sole asset or property of the business financed by you. The following transactions take place during the month of September.

Sept.		£
1	Arranged to pay a rent of £25 for the use of a small lock-up shop, and paid the landlord a month's rent in advance	25.00
3	Bought miscellaneous goods for sale	75.00
	Purchased stamps from Post Office	5.00
	Cash sales	55.74
6	Hired a part-time assistant and paid wages	30.00
	Paid for advertising in local paper	12.20
8	Cash sales	64.82
	Bought further goods for sale	45.40
	Paid carriage on purchases	4.50
10	Cash purchases	15.66
	Withdrew cash for household expenses	50.00
14	Paid wages	30.00
	Cash sales	54.30
18	Paid insurance premium for fire risk	16.60
	Cash sales	68.78
20	Paid wages	30.00
	Bought invoice stationery etc.	9.98
25	Cash takings for past three days	82.20
	Further purchases for stock	26.50
28	Paid wages	30.00
30	Cash withdrawn to pay private bills	50.00

Enter up these cash receipts and payments in your cash book for September, post up the corresponding double entry to the ledger accounts and take out a trial balance as at 30 September.

7

Gross profit and stock

In this chapter you will learn:
* **how to account for stock (also known as inventory)**
* **how to calculate cost of sales**

Say you have 10 articles which have cost you £1 each. You buy 20 more, also at £1, and sell 25 articles at £1.40 each. What profit have you made? You might say £5 profit (the excess of your sales receipts £35 over the total cost of the available stock you had for sale).

But what about the stock of articles not yet sold? This stock presumably has a very definite value and must be taken into account in working out your profit. It is valued at its cost price of £5 (i.e. five articles not yet sold which you bought at £1 each).

The true profit is worked out like this:

	£	£
Total sales (25 articles at £1.40 each)		35.00
Add stock unsold (5 valued at £1 cost)		5.00
		40.00
Less stock at start (10 articles at £1)	10.00	
and additional purchases (20 at £1)	20.00	10.00
Showing a final profit of		10.00

Another illustration is now shown, both arithmetically and in a proper accounting manner, leading on to the business trading account and gross profit.

A small trader, on 1 April, has a stock of 200 articles bought at the average cost of £1 each. Early in April he buys 100 more of these articles, now priced at £1.10 each. During the month of April he sells 240 articles at £1.50 each.

The trader's true profit is first worked out arithmetically in this way:

	£	£
Total sales (240 at £1.50)		360.00
Add stock unsold (60 at cost of £1.10)		66.00
		426.00
Less stock at start (200 at £1)	200.00	
plus purchases during April		
(100 at £1.10)	110.00	310.00
To show a trading profit of		116.00

In accounting, the money values of this arithmetical problem would be set out in this way:

Trading Account for the month ended 30 April

	£		£
Opening stock 1 April	200	Net sales	360
Purchases for month	110	Closing stock 30 April	66
Gross profit	116		
	426		426

However, a more useful way of setting it out is to add the opening stock and deduct the closing stock all on the debit side of the trading account, leaving only sales on the credit side. This has the advantage of calculating and clearly disclosing the cost of the goods sold – the 'cost of sales'.

Trading Account for the month ended 30 April

	£		£
Stock 1 April	200	Net sales	360
Add purchases	110		
	310		
Less stock 30 April	66		
Cost of sales	244		
Gross profit	116		
	360		360

Note that the gross profit is unchanged by the variation in style. The first account is included here simply because it is easier to follow the double entries.

....
Insight

The Gross Profit is the difference between the selling price and the purchase price, and must be large enough to cover all the business costs in order to leave a net profit.
....

Cost of sales

The cost of the goods sold (commonly called the **cost of sales**) is found by adding purchases to the opening stock, and then deducting the closing stock at its valuation.

The purpose of the trading account is to find the main or gross profit (or loss) on trading for a certain stated period. The gross profit is simply the balance of the account, the difference between the net sales and the cost of sales.

Valuation of stock

In the commercial world, the valuation of the stock not yet sold
at the close of an accounting period is taken to be either at its cost
price (the price at which it was bought) or at the current market
price, *whichever is the lower*. When working elementary exercises,
students are given the closing stock figure as a rule, unless the
problem is specifically based upon the calculation of stock. Even
so, establishing the correct value is not always easy. If stocks
cannot be identified to a particular purchase, and are simply kept
in bulk, what cost price is used? Is it assumed that the first stocks
bought were the first ones to be used, or that the last bought were
the first used? In practice, the former (known as **FIFO** – first in,
first out) is normally used, rather than the latter (known as
LIFO – last in, first out).

The stock account

The opening stock of an established business is already a debit
balance in the books at the beginning of a new accounting period.
It will have been debited to the stock account at its valuation at
the close of the previous accounting period. Consequently the *old
stock figure* of an established business is taken to the debit of the
trial balance when final accounts are being prepared for the current
trading period.

In examination papers, the new closing stock figure can always be
identified by its date. Generally, the closing stock is brought into
the books *after the trial balance stage*, by a new debit to the stock

account and with its corresponding credit to the newly prepared trading account.

Note how the stock account on the following page is made up from the information given earlier in this chapter. The opening stock (£200) at 1 April will be *transferred* from the debit of stock account to the debit of the trading account. This transfer is shown on the credit of the stock account, balancing this part of the old account as at 30 April. At the same time, the £66 debit for the more recent valuation of closing stock at 30 April is now brought into the books for the first time by *creating a new debit* on this account and with its corresponding double entry being credited to the trading account, or, alternatively being deducted on the debit side of the trading account.

Stock Account

Dr.						Cr.
Apl.		£	Apl.			£
1	Opening stock at beginning of month	200	30	Transfer of old stock to trading account		200
30	Trading account (closing stock now brought into the books)	66				

- ▶ Items of *prime cost* will be found on the debit of the trading account: the opening and closing stock figures, the net purchases, and the wages of production or of preparing the goods for sale. Also debited to the trading account will be expense items which vary with the turnover, such as delivery charges.
- ▶ On the credit of the trading account will be found the net sales.

TESTING YOURSELF

7.1 Make up the trading account of Edwina Ewing, showing her cost of sales figure from the information given below.

May

At the beginning of the month there was a stock of 1800 items, bought at the average cost of £1 each. The total net purchases for May were 3400 items bought at an average cost of £1.25.
Total net sales of 3800 articles realized £7300 from cash customers.

Ascertain Miss Ewing's final stock figure, place a valuation on it, and make up her trading account for May.

7.2 Extract the information you need from the trial balance below, to make up the trading account of Miss Vivette Green for the month of October.

Trial Balance of Vivette Green

31 October	£	£
Capital account 1 Oct.		2500
Drawings for month	180	
Net purchases	4800	
Net sales		6800
Carriage on purchases	50	
Warehousing expenses	220	
Stock 1 Oct.	1150	
Cash balance 31 Oct.	2900	
	9300	9300
Miss Green valued her closing stock on 31 October at £1550.		

7.3 What is the cost of sales figure in the preceding problem? Explain why, although recent purchases comprise much of the closing stock valuation, the figure for net purchases differs from the cost of goods sold.

8

Trading and profit and loss accounts

In this chapter you will learn:
- *the difference between gross profit and net profit*
- *how to prepare trading and profit and loss accounts*
- *about carriage inwards and carriage outwards*

In the preparation of the final accounts, the profit element is split, for convenience, into two parts:

▶ the **gross profit,** *already described as the difference between the cost of sales and the actual sales total*
▶ the **net profit,** *which is the final trading profit of a business, after all office, general and distribution expenses have been charged against the gross profit brought down from trading account.*

Illustration by arithmetic

Another arithmetical example is shown. Say a trader buys £500 of goods for sale. She pays £10 delivery costs and sells the whole consignment for £780, after incurring office and distribution expenses of £120. Arithmetically, her gross and net profit would be worked out in this way:

		£
Net sales on the whole consignment		780
Less purchases and	£500	
delivery costs	10	510
To give a gross profit of		270
Less office and distribution expenses		120
To show a final net profit of		150

In this illustration the opening and closing stocks have been
omitted for the sake of simplicity.

Making up the final accounts

We will now look back at the trial balance in Chapter 6 and make
up a trading and profit and loss account from those details which
relate specifically to the revenue income or expenditure of the
business, namely purchases, sales, advertising and salaries.

Since this business was started from scratch, there is no opening
stock, but it would be reasonable to assume that some stock
would be unsold at the end of the first month's trading. If we place
a valuation on the stock of purchases not yet sold of £800, the
trading and profit and loss account will be made up on these lines.

Trading and Profit and Loss Account for the month ended 30 June

	£		£
Stock 1 June	nil	Sales for month	2125
Purchases for month	1600	Stock 30 June	800
Gross profit c/d	1325		
	2925		2925
Advertising	73	Gross profit b/d	1325
Salaries	600		
Net profit	652		
	1325		1325

The trading and profit and loss account, as its name suggests, is really two accounts put together. At the top is the trading account, showing the calculation of gross profit after deducting the cost of sales. This gross profit is then taken down as the opening credit balance in the profit and loss account. It is rare for further income to be added on the credit side; instead the profit and loss account shows all the other expenses (office, marketing, etc.) that have to be deducted in order to arrive at the final net profit or net loss. A net profit is a credit balance on the profit and loss account; a net loss is a debit balance.

Although the account set out in this way is easy to follow, it is invariably set out in vertical format in modern financial statements, looking like this:

	£	£
Sales per month		2125
Less cost of sales:		
Purchases per month	1600	
Plus opening stock	–	
Less closing stock	800	
		800
Gross profit		1325
Advertising	73	
Salaries	600	
		673
Net profit		652

Business revenue and expense

The only figures extracted from the trial balance in Chapter 6 to make up the revenue account were those affecting the firm's trading profits. Three important balances were excluded in working out the business profit, namely capital, drawings and cash. A simple illustration will explain this:

Suppose you have £10 in your pocket. You buy £8 worth of goods, incur selling expenses of £3, and then dispose of your entire stock for £15. Your profit is £4 (£15 sales less £8 purchases plus £3 expenses).

Total sales		£15	
Less purchases	£8		
and expenses	3	11	= £4 profit

Now if you consider this problem carefully, it will be seen that your original capital of £10 has not figured in your profit calculation, and neither has your final cash balance entered into it.

Also, if you had withdrawn £1 cash to buy your small sister some sweets, this sum would certainly have reduced your cash balance, but it would have *no effect upon the profit already made*.

Two kinds of carriage

Carriage, i.e. delivery charges, on purchases (**carriage inwards**) is an additional expense on the goods bought, increasing the cost of sales, and is a trading account expense.

Carriage on sales (**carriage outwards**) is a selling expense and is debited to the profit and loss section along with the other administration and distribution expenses.

Note that both aspects of carriage are debits, and must not be confused with returns inwards and returns outwards explained in Chapter 16.

Production wages and office salaries

The wages of production or of warehousing vary with the turnover of the business, and are debited to the trading account as part of the cost of the goods produced.

The salaries of the office and administrative staff are charged to the profit and loss account. Like other routine expenses such as rent, rates and insurance, salaries are a more constant figure and not so much affected by short-term variation of production volume.

On no account must the amounts spent on the purchase of fixed assets and business property such as machinery, motor vehicles, fixtures and equipment be debited to either of these revenue accounts. Later, though, it will be explained that a kind of 'wear and tear' allowance called **depreciation** is included in the accounts.

Credits on the revenue accounts

Few items appear on the credit of either the trading or the profit and loss account. Generally, when they do, they often refer to small and miscellaneous profits and are usually taken to the profit and loss account, and are identified by the words 'received' or 'recovered' as in the instances of discounts received, commissions received, rents received and bad debts recovered.

Bringing closing stocks into the books

When the new valuation figure for closing stock is credited to the trading account, a corresponding debit must be made on the stock account to provide a double entry.

Since we are now reaching the final stage of the illustration from Chapter 3, and the closing stock at 30 June has been valued at £800 and already taken to the credit of the trading account at the beginning of this chapter, we must now open a new stock account and debit this same amount to complete the double entry.

Stock Account (20)

Dr.   Cr.

June 30	Trading account		£ 800				

PAUSE FOR THOUGHT

▶ *What is meant by the terms: revenue accounts, cost of sales, gross profit and net profit?*
▶ *Why is the main revenue account of the business split into two parts or sections?*

Insight

Carriage inwards (on purchases, i.e. goods coming in) joins the purchase debit in the trading account, whereas **carriage outwards** (on sales, i.e. goods sent out) is a selling and delivery expense to be debited to the profit and loss account. *Both are debit entries.*

Neither assets (other than stocks) nor capital expenditure appear in revenue accounts. They only appear in the Balance Sheet.

Remember, the revenue accounts are used to match operating expenditure with the revenue made in the same period.

KEY POINTS

▶ The heading of a revenue account should show the period covered and the date to which the account is made up.
▶ Unless stated to the contrary, production wages are normally taken to the trading account, and office salaries to the profit and loss account. Some discretion should be used, as wages

(Contd)

paid to office cleaners are chargeable to the profit and loss account and the production manager's salary is chargeable to the trading account.

▶ In the beginners' trial balances opening stock is a debit, but closing stock (only just valued at the end of the period) *does not become a book entry until it is credited first to the trading account*.

▶ Occasionally a business sustains a net loss. The effect on the trader's capital is explained in the next chapter.

TESTING YOURSELF

8.1 Re-draft the incorrect statement below to show clearly the correct gross and net profit of this business.

Trading and P+L Account of Annabel Royle for December

	£		£
Stock 31 Dec.	300	Sales	6200
Purchases	2850	Carriage	
Salaries	200	outwards	30
Gross profit c/d	3300	Stock 1 Dec.	420
	6650		6650
Carr. inwards	60	Gross profit	
Warehouse wages	1600	b/d	3300
Travelling	45		
Insurance	35		
Rent	80		
Drawings	100		
Net profit	1380		
	3300		3300

8.2 Extract the necessary information from this trial balance and make up the trading and profit and loss account of Jasinder Singh, master plumber, for the month of June.

Trial Balance 30 June

	£	£
Capital W.B. 1 June		6 000
Drawings for month	400	
Van at valuation	1 750	
Office fittings at cost	350	
Purchases and sales	5 200	8 800
Carriage inwards	52	
Carriage outwards	46	
Advertising	108	
Insurance	30	
Petrol/oil	24	
Heating/lighting (workshop)	34	
Heating/lighting (office)	14	
Wages (workshop)	1 444	
Office salaries	420	
Stamps	8	
Stock 1 June	1 850	
Cash balance 30 June	3 070	
	14 800	14 800
Stock valuation at 30 June £2350		

8.3 Explain the following terms:

a *carriage inwards*
b *carriage outwards*
c *drawings.*

9

The balance sheet

In this chapter you will learn:
- **what a balance sheet is**
- **how to close the remaining ledger accounts**

The balance sheet takes care of the remaining 'open' balances in the books. After incorporating the retained profit (or loss) on trading, it shows the assets, capital and liabilities on a certain date.

The balance sheet is *not an account* but a financial statement, grouping and listing the business property and the capital and liabilities on that certain date.

Methods of presentation

This is the final stage of the illustration commenced in Chapter 3. First shown in the modern vertical style, which is now universally adopted, it would be presented in this way:

Balance Sheet as at 30 June

Assets employed	£	£
Stock on hand 30 June	800	
Cash in hand	1352	2152
Financed by		
Proprietor's capital 1 June	2000	
Retained profit	152	2152

'Retained profit' is that balance of profit remaining after all appropriations (or withdrawals in the case of a sole trader) have been deducted.

If this top half of the vertical statement is tilted over to the right, the older horizontal 'T' style balance sheet is shown. Although this is no longer used, it is easier to follow double entry transactions in the horizontal presentations.

Balance Sheet as at 30 June

Capital and liabilities	£	£	Property and assets	£	£
Capital 1 June	2000		Stock 30 June	800	
Add net profit	652		Cash in hand	1352	2152
	2652				
Less drawings	500	2152			
		2152			2152

Note that in horizontal presentation the proprietor's capital account is generally shown in fuller detail. In effect, it becomes a vertical copy of the owner's own personal ledger account (see page 62).

In this simple illustration there are only two assets and one liability at the balancing date, but even this small balance sheet has a story to tell. You started in business a month ago, made a small trading profit of £652, withdrew £500 on account of profits (thereby relinquishing part of your capital holding), and finally finished up with £152 more in asset value than when you started. Your capital or net worth at the end of the month of June is represented by the two assets, stock or merchandise not yet sold valued at £800, and £1352 cash in hand; your net assets have thus increased by £152 since the beginning of June.

Net loss on trading

Alternatively a net loss might result through adverse or unfortunate trading conditions. The *debit balance* on the profit and loss account would be transferred to the *debit* of the proprietor's capital account in addition to personal drawings for the period. The net loss on trading would be reflected on the assets side of the balance sheet by a corresponding decrease in total net assets.

Capital and drawings

The capital account of the sole trader is an internal obligation of the business towards its owner, and must be kept quite distinct from the outside liabilities (trade and expense creditors to be introduced a little later). The business capital, now represented in the form of stock, continues to finance the day-to-day trading operations of the business, but any withdrawals of money (or goods) for the private and family use of the proprietor involve a corresponding reduction of the capital account. Bear in mind, though, that we are in no way concerned how the owner of a business uses the cash or the goods withdrawn in this way.

Closing of ledger accounts

In the exercise just completed, the proprietor's capital account is brought up to date by transferring the net profit (a credit balance of excess sales revenue) to the credit of that account, and also by transferring the debit balance on drawings account to the debit of capital account, as now shown:

Drawings Account (4)

Dr.							Cr.
June			£	June			£
30	Cash	CB1	500	30	Transfer to capital account	2	500

Capital Account (2)

Dr.							Cr.
June			£	June			£
30	Transfer from drawings account	4	500	1	Cash	CB1	2000
30	Balance c/d		2152		Net trading profit for month		652
			2652				2652
				July			
				1	Balance b/d		2152

The recently created stock account remains open with its new debit balance of £800 at 30 June, but the remainder of the old revenue accounts is now *closed by transfer* to either the trading or profit and loss account.

Sales Account (15)

Dr.				Cr.		
June		£	June			£
30	Transfer to trading account	2125	5	Cash	CB1	458
			18	Cash	CB1	562
			24	Cash	CB1	1105
		2125				2125

Purchases Account (5)

Dr.							Cr.
June			£	June			£
2	Cash	CB1	750	30	Transfer to		
20	Cash	CB1	850		trading account		1600
			1600				1600

Advertising Account (8)

Dr.							Cr.
June			£	June			£
12	Cash	CB1	73	30	Transfer to		
					P&L a/c		73

Salaries Account (9)

Dr.							Cr.
June			£	June			£
30	Cash	CB1	600	30	Transfer to		
					P&L a/c		600

Insight

Remember, as mentioned before, a balance sheet is only a snapshot in time looking at past events, i.e. transactions that have already taken place.

An accounting period of 12 months is generally used for financial reporting.

KEY POINTS

▶ Testing yourself 9.1 is simple arithmetic. Remember, though, to add back drawings to the difference in capital values between the beginning and the end of the

(Contd)

trading period. The sum withdrawn is part of the true profit made.

▶ One main feature about vertical presentation of the balance sheet is that the figures take on more positive and realistic meaning, with the capital shown as representing the holding and value of net assets. The vertical style balance sheet has superseded the old horizontal 'T' style in all forms of trading and professional firms in addition to limited companies.

The vertical style trading and profit and loss account has also become an accepted feature with most large organizations. For the time being, however, the revenue accounts will be presented in the older and more closely related two-sided accounts, probably easier to comprehend from a beginner's viewpoint with the similarity to the double entry ledger system itself.

▶ In elementary bookkeeping examinations, although arithmetical accuracy is important, even more important is the correct grouping and presentation of the figures, in particular on the balance sheet.

TESTING YOURSELF

9.1 Davidia Kneale started in business a year ago with £5000 cash as capital. Her present capital, at 31 December, after one year's trading, cash transactions only, is represented by:

	£
Furniture/fittings	840
Stock of goods unsold	2680
Money in hand	3150

Miss Kneale has withdrawn £3180 on account of profits during the year. Can you work out her trading profit?

9.2 You are required to assemble Li Kwan Chung's ledger balances (shown below) in trial balance form, with a view to finding out his opening capital. In making out his final accounts for the month ended 31 May, show the cost of sales figure on his trading account.

	£
Stock 1 May	740
Purchases	3330
Sales	4650
Rent/rates proportion	136
Drawings	200
Salaries	400
Cash balance 31 May	1780
Insurance	12
Warehousing expense	30
Carriage inwards	7
Advertising	15

Mr Li valued his closing stock at £530.

9.3 Explain the purpose of a Balance Sheet.

9.4 What is retained profit?

10

Illustrative example

In this chapter you will revise what you have learned so far.

As a basis for revision we take over the final balances shown on the balance sheet at 30 June in Chapter 9, and continue with a few further cash transactions for the month of July.

This means that we start the new period with the two debit balances of cash in hand £1352 and stock £800, and one credit balance of £2152, now our opening capital at 1 July.

As from the beginning of this new period our total debits equal our total credits: $A = C + L$, where A is the symbol for the assets, C for the capital and L for the liabilities.

The trading transactions for the month of July are as follows.

July		£
2	Bought further goods to be re-sold	751
	Carriage on purchases	50
4	Cash sales	375
8	Paid for stationery	86
9	Cash sales	1182
14	Bought new counter (see text below)	750
15	Received cash for goods sold	823
24	Cash sales	1357
	Carriage on sales	20
	Paid for further purchases	540

25	Withdrew for family expenses	1250
28	Paid salary to part-time assistant	860
30	Postages bought during month	80

The closing stock at 31 July was valued at £380.

Note that the transaction on 14 July is a special purchase of a *fixed asset*, bought for permanent use in the business and not for re-sale. The cost of this asset (£750) will be debited to a separate ledger account called the fittings account, to appear ultimately under its own heading on the assets side on the balance sheet at 31 July.

The various accounts are now written up.

Cash Account (2)

Dr.   Cr.

July			£	July			£
1	Balance brought forward from June	CB1	1352	2	Purchases	6	751
					Carriage inwards	17	50
4	Cash sales	16	375	8	Stationery	22	86
9	Cash sales	16	1182	14	Fittings	3	750
15	Cash sales	16	823	24	Carriage outwards	18	20
24	Cash sales	16	1357		Purchases	6	540
				25	Drawings	4	1250
				28	Salaries	9	860
				30	Postages	24	80
				31	Balance c/d		702
			5089				5089
Aug.							
1	Balance b/d		702				

It will be seen that the balance of cash in hand on 1 July is *only part of the business capital*, the other part being invested in stock, as now shown by the stock account as at 30 June brought forward.

Stock Account (20)

Dr.   Cr.

June 30	Trading account		£ 800	June			£

The capital account is also brought forward to show clearly all the ledger accounts as from the beginning of the new period.

Capital Account (2)

Dr   Cr.

				June 1	Balance b/f		£ 2152

The remainder of the ledger accounts for the new period, to the completion stage of the double entry from the cash account, are as follows.

Purchases Account (6)

Dr.   Cr.

July 2	Cash	CB2	£ 751	July			£
24	Cash	CB2	540				

Sales Account (16)

Dr.   Cr.

				July 4	Cash	CB2	£ 375
				9	Cash	CB2	1182
				15	Cash	CB2	823
				24	Cash	CB2	1357
							3737

Carriage Inwards (17)

Dr.   Cr.

July			£				
2	Cash	CB2	50				

Stationery (22)

Dr.   Cr.

July			£				
8	Cash	CB2	86				

Carriage Outwards (18)

Dr.   Cr.

July			£				
24	Cash	CB2	20				

Salaries (9)

Dr.   Cr.

July			£				
28	Cash	CB2	860				

Postages (24)

Dr.   Cr.

July			£				
30	Cash	CB2	80				

Drawings (24)

Dr.   Cr.

July			£				
25	Cash	CB2	1250				

Fittings Account (3)

Dr.   Cr.

July			£				
14	Cash	CB2	750				

The postings have now been completed for the month of July. A trial balance will be taken at this stage to prove the accuracy of the postings.

Trial Balance 31 July

		£	£
Cash in hand 31 July		702	
Capital 1 July			2152
Drawings		1250	
Fittings		750	
Purchases		1291	
Sales			3737
Carriage inwards		50	
Carriage outwards		20	
Stationery		86	
Salaries		860	
Postages		80	
Stock 1 July		800	
		5889	5889

Note that the opening stock at 1 July is debited to this general form of trial balance, but that the *closing stock of £380 at 31 July is not yet a balance in the books until the trading account is made up.*

All the revenue accounts for July are now closed (sales, purchases carriage inwards and outwards, stationery, salaries and postages) by transfer of their balances either to the trading account or to the profit and loss section of this combined revenue account.

Trading and Profit and Loss Account for the month ended 31 July

	£	£	
		£	
Stock 1 July	800	Sales for July	3737
Purchases	1291		
Carriage inwards	50		
	2141		
Less stock 31 July	380		
Cost of sales	1761		
Gross profit c/d	1976		
	3737		3737
Carriage outwards	20	Gross profit b/d	1976
Stationery	86		
Salaries	860		
Postages	80		
Net profit for July	930		
	1976		1976

The opening stock at 1 July (£800) has now served its purpose by transfer of its debit balance to the trading account. That part of the stock account may now be ruled off. At this stage, too, the new valuation for closing at 31 July (£380) is *brought into the books for the first time by a credit to the trading account* (shown as a deduction on the debit side). The double entry and corresponding debit to stock account is now shown:

Stock Account (2)

Dr.   Cr.

		£			£
June 30	Trading account	800	July 31	Transfer of old stock to debit of trading a/c	800
July 31	Trading a/c (new stock brought into books)	380			

At this final stage the assets are listed on the balance sheet in order of permanency, fittings being regarded as more permanent than stock, and stock thought to be rather more permanent than cash, with which it has been bought. This is simply a question of preference, generally followed by the majority of industrial, commercial and professional organizations.

The balance sheet is now drawn up for the second month of the illustration started in Chapter 3. Note that the proprietor's equity and capital holding has decreased since 1 July by £320, the amount of the withdrawals in excess of the profit made during the month of July.

Balance Sheet as at 31 July

Assets employed		
Fixed assets	£	£
Fittings		750
Current assets		
Stock at 31 July	380	
Cash in hand	702	1082
		1832
Financed by		
Capital 1 July	2152	
Add net profit	930	
	3082	
Less drawings	1250	1832
		1832

CHECKPOINTS

Before starting Testing yourself 10.1, check how you are doing on the following routine which, by this stage, should be becoming automatic.

1 Are your worked exercises reasonably neat and tidy, with the correct alignment of figures in the total columns?

2 Occasionally mistakes will occur. Instead of scratching out, simply rule through the error and insert the correct amount over the top.

3 Never rule free-hand. Always use a good ruler and keep any corresponding totals on the same horizontal level.

4 Are your headings clear and bold with a space for neatness between the ledger account heading and the details within the account?

5 Do you confine the use of the £ sign to the top of the amount column only, and write in the name of the month only at the top of the date column?

TESTING YOURSELF

10.1 You start a part-time business on 1 January with a capital of £1500 comprising these assets:

	£
Second-hand fittings (counters and shelving) valued at	500
Stock of goods bought at the special price of	200
Available cash for business use	800
	1500

First record these details in your financial book as follows.

Debit £500 to fittings account ⎫ and credit £1500
Debit £200 to stock account ⎬ to your opening
Debit £800 to cash account ⎭ capital account.

Note that these three assets totalling £1500 make up your *opening capital* in this instance. *The £800 cash is only part of your capital.*

After recording the opening entries, open the cash book and the ledger and post up the following transactions for the month of January. Take a trial balance at 31 January, and make up the trading and profit and loss account and a balance sheet as at 31 January. The valuation of your closing stock is £375.

Jan.		£
2	Bought a second-hand van. Paid cash (debit van account – credit cash)	650
3	Cash sales	162
	Bought stationery	8
	Paid rent for month	35
5	Bought goods for re-sale	188
	Cash takings	154
8	Paid fire/theft insurance (monthly payment)	25
	Cash sales	107
	Drew for 'self'	50

10	Paid part-time wages for warehousing	30
	(see note on wages in Chapter 8)	
	Travelling expenses of staff	14
	Bought a second-hand filing cabinet for office	85
14	Cash sales	76
16	Cash sales	166
	Paid wages	30
22	Paid for advertising	27
	Bought postage stamps	8
	Paid wages	30
23	Cash sales	148
	Further purchases to replenish stocks	155
	Paid carriage on goods bought	12
25	Cash sales	78
	Paid petrol account	22
	Repairs to van	24
	Cash purchases	186
28	Drew for 'self'	50
	Paid wages	30
31	Cash sales	72

Show as two distinct accounts, van account and fittings account, and debit the purchase of the filing cabinet (£85) to the latter.

These *fixed asset* accounts are shown under their own headings at the top of the assets on the balance sheet.

11

Cash and bank transactions

In this chapter you will learn:
- *how to separate cash and bank transactions*
- *how to keep a two column cash book*

All transactions up to this stage have been on a cash basis. Money has been received for cash sales, and all purchases and expenses of the business have been paid in cash, i.e. in notes and coin. Only a single-column type of cash account has been needed.

This may suit a small shop-keeper with sales across the counter for cash, but in most businesses these days there is a certain amount of credit dealing, and settlement by cheque or credit transfer has become both necessary and convenient. For simplicity at this stage it is assumed that all payments are made by cheque.

The bank account

It has been emphasized that the cash account is part of the ledger system. It still remains an integral part of the old system, but when dealing with a bank account, a specially ruled ledger, called the **cash book**, is used instead.

The cash book has additional columns for bank transactions, and the two accounts for *both cash and bank* are balanced up independently. The account columns for cash and bank (cheques) are simply kept *side by side* for convenience, as quite often money (both cash and cheques) is paid out of the office cash surplus into the bank, and vice versa cash is withdrawn from the bank by cheque for the needs of the office.

Generally, both accounts have debit balances, but the bank account, unlike the cash account, could have a credit balance if the money at the bank was overdrawn. It would then be referred to as an **overdraft**.

In these early cash and bank exercises it is usual for the student to assume that cheques from customers are banked on the same day that they are received. In practice, however, this may vary.

An illustration is now shown of the two-column cash book. Note that the cash and bank columns are two separate accounts and are balanced up independently of each other. The bank columns in these early exercises relate entirely (on the debit side) to cheques received from customers and surplus cash paid into the bank and (on the credit side) to cheques drawn by the firm, withdrawing cash from the bank or in settlement of creditors' accounts.

Two-column Cash Book

Date		Cash	Bank	Date		Cash	Bank
Oct.		£	£	Oct.		£	£
1 Balance b/f		50	2000	2 Contra	c		200
2 Contra	c	200		4 Rent			300
8 Cash sales		628		6 Purchases			885
12 L. Harris		227		Cleaning		125	
18 Cash sales		1102		9 Stationery		48	
Contra	c		1000	Stamps		52	
22 P. Marsden Ltd			1800	12 Purchases			666
28 Cash sales		825		Advertising			84
30 S. Roberts			789	16 T. Jones			423
31 Contra	c		1500	18 Contra	c	1000	
				19 Petrol/oil		82	
				24 Cleaning		125	

(Contd)

					28	G. Rawson & Son			555
					30	Salaries			1850
					31	Contra	c	1500	
						Drawings			550
						Balance c/d		100	1576
			3032	7089				3032	7089
Nov. 1	Balances b/d		100	1576					

The small letter 'c' in the folio columns indicates contra entries, or transfers between office cash and bank and vice versa. These postings, shown on both sides of the cash book, debit in one column for the corresponding credit on the other side, cancel themselves out, so that *ledger postings are unnecessary*.

Any charges made by the bank (interest on overdrafts etc.) are shown as deductions on the bank statement sent to the customer at the end of the month, or sometimes quarterly. The firm's book-keeper, on checking the bank statement against the company's cash book bank columns, will enter any bank charges of this nature on the *credit side* of the cash book bank column. He or she will complete the double entry by posting the amount to the *debit* of the bank charges account in the nominal or expense ledger.

Cash or cheque?

Sometimes, in an examination question, it may not be clear whether a receipt or a payment is to be entered in the cash or the bank column. In the absence of instructions, treat all payments from customers, and all payments to suppliers, as *cheque payments*, since cash would not be sent through the post. On the other hand, the purchase of a pint of milk for the coffee, payment of a few pounds to a window cleaner, etc. are probably made in cash.

Primary records

The entries on the debit side of the cash book come from the daily till rolls of cash registers, duplicate receipt books and from the listed amounts on the bank paying-in slips.

Entries on the credit side of the cash book originate from the official receipts from suppliers, cash memos from sundry small purchases, and the counterfoil stubs of the firm's cheque books.

Insight

The important thing is to avoid the complication of an overdrawn cash balance which, strictly, should not be possible. When in doubt, pay by cheque. Also, it is not necessary to record the names of suppliers or customers where goods are bought or sold for cash or cheque payment at the time the goods are handed over in completion of the transaction, i.e. cash sales and cash purchases transactions.

KEY POINTS

▶ Check the contra items to ensure that there is a debit to correspond with every credit. Do not guess with regard to contras. Think which account has increased or received benefit (cash or bank?). Debit cash and credit bank or vice versa as the case may be.

▶ Generally there are more items and lengthier columns to add up in the firm's cash book. The importance of correct alignment cannot be over-emphasized.

▶ Usually there are more entries on the credit or payments side of the cash book. This means leaving a few blank spaces on the debit, when balancing up, as the totals must be on the same horizontal. Remember, too, that allowance must be made on the credit side of the cash columns (and generally bank columns too) for the cash and bank balances.

(Contd)

► Where personal names are given (the names of people and firms), these names should be recorded in the details column of the cash book. Generally the receipt or payment refers to the settlement of a credit account already posted up from an original invoice to a customer's or supplier's ledger account.

TESTING YOURSELF

11.1 Enter the following transactions in a two-column cash book for the month of April, balancing up both cash and bank columns at 30 April.

April		£
1	Cash in office	400
	Cash at bank	5600
3	Cash sales	562
	Paid cash into bank	500
	Paid rent by cheque	350
8	Paid Jim Lord by cheque	284
	Cash purchases	127
14	Cheque from R. Davies	256
	Cash sales	548
	Cash paid into bank	500
18	Bought stamps for cash	30
	Drew cheque for 'self'	600
25	Paid Ossie Pye by cheque	488
30	Invoiced for shelving from A.B.C Services Ltd	363

11.2 On 1 September Charmain Wade's cash and bank balances were £100 and £6400. Her transactions during the month are listed below. Write up her cash book for September, bringing down her cash and bank balances as at 1 October.

Sept.		£
1	Cheque drawn for office cash	400
3	Bought goods from Bob West	800
5	Paid Sam Ronson's account	242
	Paid office cleaner (cash)	180
8	Cash sales	788
	Paid cash into bank	500
12	Bought stamps for cash	30
	Paid rates by cheque	1355
15	Cash sales	687
18	Cash sales	723
	Paid wages out of cash	550

(Contd)

	Drew cheque for 'self'	800
22	Paid Josh Brown's account	364
	Leslie Frost paid by cheque	684
24	Cash sales	949
	Paid cash into bank	500
26	Cash purchases	152
	Bought stamps for cash	50
27	Paid advertising account	225
29	Sarah Raven sent cheque in payment of old account	250
30	Miss Wade paid all surplus cash over £100 into the bank	

11.3 Explain the following terms:

    a cash book
    b overdraft
    c contra entry.

11.4 Why should you avoid having an overdrawn cash balance?

11.5 Where do the entries on the credit side of the cash book originate from?

12

Bank reconciliation

In this chapter you will learn:
- *how to check the bank statement*
- *how to reconcile it to the cash book*

In this chapter we concern ourselves solely with the bank columns of the cash book and the agreement or reconciliation of these columns with the bank's own statement forwarded monthly, or to bank customers on request.

Normally the bank balance according to the cash book does not agree with the balance according to the bank's own statement on the same date. This is due to the time lag between the receipt of cheques from the firm's customers and their clearing from the bank accounts of the debtors. Again, there is a similar delay between the handing over or posting of cheques to suppliers or creditors and their collection, via their bankers, from the account of the paying firm.

Checking the bank statement

Apart from the delay in clearing, presenting and collection of cheques, there are often other bank debits for charges and expenses that appear periodically on the bank statement. Instances of these are the interest charged by the bank on loans and overdrafts, and sometimes charges per cheque drawn. Often there will be payments out that have been made as direct debits or standing orders in compliance with the instructions of the customer. Equally, the business may have received standing orders or direct debits. It is also increasingly common for bills to be met by direct transfer between bank accounts using the BACS or CHAPS systems.

Dividends from investments owned by the business are often paid direct to the bank, and it is only when credited to the firm's current account and checked through reconciliation by the cashier that these amounts will be entered in the cash book. The cash book in this way is brought up to date as far as possible before the final reconciliation takes place, which involves ticking, item for item, all the debits in the cash book against all the credits on the bank statement, and all the credits in the cash book against all the debits on the bank statement.

A note is made of *all amounts remaining unticked* in either the cash book or the bank statement. It is these unticked items which form the basis for the reconciliation.

Cash Book (bank columns only)

Dr.   Cr.

June		Cash £	Bank £	June		Cash £	Bank £
1	Balance b/f	✓	3000	2	Cheque to X	✓	162
8	Cash paid in	✓	702	5	Rates	✓	440
10	Cheque from A	✓	245	12	Wages	✓	600
15	Cheque from B	✓	363	15	Advertising	✓	85
25	Cash banked	✓	661	24	Wages	✓	600
28	Cheque from C		555	28	Cheque to Y		97
				30	Cheque to Z		364
					Balance c/d		3178
			5526				5526
July							
1	Balance b/d		3178				

Bank Statement

June		Debit £	Credit £	Balance £
1	Balance b/f			✓ 3000
4	X	✓ 162		2838
8	Cash receipts		✓ 702	3540
9	Rates	✓ 440		3100
12	Wages	✓ 600		2500
	A		✓ 245	2745
18	B		✓ 363	3108
19	Chq. no. 22643	✓ 85		3023
24	Wages	✓ 600		2423
25	Cash receipts		✓ 661	3084
28	TS dividend		500	3584
	Standing order	105		3479
30	Charges	40		3439

Procedure for reconciliation

Note that all items in the firm's cash book have been ticked off with the exception of:

▶ C's cheque for £555 not yet cleared by the bank
▶ the cheques made out to Y and Z for £97 and £364 not yet presented for payment.

All items on the bank statement have been ticked off except for the debits against the firm for £105 and £40, and the dividend of £500 paid direct to the bank.

The firm's bookkeeper now brings the company cash book up to date as far as possible, to show an *amended bank balance* of £3533 thus:

Bank Reconciliation Statement

Dr.   Cr.

June				£	June				£
30	Balance b/f			3178	30	Standing order			105
	Treasury stock div.					Bank charges			40
	Paid direct to					Balance c/d			3533
	bank			500					
				3678					3678
July									
1	Balance b/d			3533					

The cheque not yet cleared and the cheques not yet presented for payment are adjusted on the reconciliation statement as follows.

Bank Reconciliation Statement 30 June

		£	£
Amended balance as per cash book			3533
Add cheques drawn but not yet			
presented for payment	Y	97	
	Z	364	461
			3994
Deduct cheque not yet cleared (i.e. not yet			
credited to the firm's current account)			555
Balance according to bank statement			3439

Insight

Note that when you pay cash and cheques into your bank account, your current account is credited, increasing your bank balance; but when you withdraw money or make a payment by cheque, your bank current account is debited, reducing your bank balance.

KEY POINTS

▶ You will probably realize, at this stage, that a 'debit' bank balance in the firm's cash book means that there is money at the bank.

From the bank's accounting aspect, however, you are *in credit*, i.e. the bank owes you money.

▶ 'Cheques not presented yet for payment' are those cheques given or sent to creditors who have not yet claimed or collected payment from your bank.

'Cheques and cash not yet cleared' refers to remittances (currency and cheques) paid into your own bank, but not yet shown on your bank statement. Often this is the total of the last paying-in slip handed in to the bank on the last day of the month.

▶ The reverse of the conventional procedure applies when dealing with a bank overdraft, your bank account being 'in the red'.

Cash and cheques paid into the account reduce the overdraft (because it is a credit balance and these are debit entries), whereas further cheques drawn increase it.

▶ The correct figure for the business bank balance to be taken to the trial balance and also to the balance sheet is always the *adjusted or amended figure* according to the cash book.

TESTING YOURSELF

12.1 The cash book bank balance of Alan Jones on 30 June shows a debit balance of £728.24, whereas the statement given to him by his bank on the same date shows a credit balance in his favour of £696.93.

Checking the bank statement against his cash book, Mr Jones finds that cash and cheques paid into his bank on 30 June, totalling £224.75, had not been credited to his current account; and two cheques for £82.64 and £110.80 sent to Holt Bros Ltd and Eric Hardwick on 28 June had not yet been presented to his bank for payment.

You are required to draw up a reconciliation statement showing the difference between the two balances mentioned in the first paragraph above.

12.2 The bank statement of Linda Owen made up to 31 December, received early in the New Year, showed her current account overdrawn by £24.54.

Checking the statement against the bank columns of her cash book, she noted the following.

- **a** *Cash and cheques totalling £178.28 paid into the bank on 31 December had not been credited to her account.*
- **b** *A cheque for £83.40 sent to her supplier on 29 December had not yet been presented for payment.*
- **c** *A dividend of £60 Treasury Stock had been credited direct to her account.*
- **d** *Bank charges of £12.50 had been debited to her account by the bank.*

What was the bank balance shown by Miss Owen's cash book on 31 December?

12.3 Why is the bank balance in the cash book normally different to the bank balance in the bank statement?

12.4 What does a 'debit' bank balance in the firm's cash book mean?

12.5 Explain the following terms:

   a *Cheques not presented yet for payment.*
   b *Cheques and cash not yet cleared.*
   c *In the red.*

13

The petty cash book

In this chapter you will learn:
* *how to keep an imprest petty cash book*
* *the source documents for petty cash entries*

Under what is known as the **imprest system** of keeping petty cash, a round sum of money (say £20) is put aside for the payment of small expenses of the office.

At the end of the week or the month, the *exact amount of money paid out* (the **total disbursements**) is added to the petty cash, so that the balance of petty cash in hand is brought back again to the original amount of the imprest, i.e. £20.

An analysis cash book, with columns specially ruled to suit the needs of the particular office and the general expense payments, is used for this purpose. This book is called the **petty cash book**.

Posting of the petty cash expense totals

The petty cash book is part of the double entry system because the amounts added to petty cash appear as a credit in the main cash book. The totals of the various expense columns of the petty cash book are extracted and debited to the relative expense accounts in the nominal ledger, and at the end of the trading period the balance of petty cash in hand is included or absorbed into the main cash balance.

Insight

Using a petty cash book stops the main cash book getting cluttered up with lots of small entries. All small miscellaneous expenses can be included here.

ILLUSTRATION

A simple style of petty cash ruling is shown below with analysis columns for four classes of routine expense. The opening balance of petty cash on hand, the imprest amount, at the beginning of the month of January is £20. The expense payments are listed for the first week of the New Year, and this page of the petty cash book balanced up on 6 January.

Disbursements for the first week of January:

Jan.		£
1	Paid bus fare	0.68
2	Bought postage stamps	2.00
	Paid for office cleaning materials	4.00
3	Bought coffee	1.64
4	Paid for pencils	0.76
	Bought more stamps	1.00
5	Taxi fare to station	2.25
	Biscuits	1.00
6	String	0.65

Note that each item of expense in the petty cash book is shown both in the total column and in the appropriate analysis column. The totals of the combined analysis columns agree with the sum of the main total column (£13.98). This is the total petty cash expenditure for the week, and, when deducted from the original imprest amount of £20.00, leaves a balance of £6.02 petty cash in hand at 6 January.

Petty Cash Book

Rec'd from cashier	Date	Details of expense	Voucher no.	Total	Stamps Stat'y	Travel	Cleaning	Sundries
£				£	£	£	£	£
20.00		Jan.						
	1	Imprest b/f						
	1	Bus fares	–	0.68		0.68		
	2	Stamps	1	2.00	2.00			
		Cleaning	2	4.00			4.00	
	3	Coffee	3	1.64				1.64
	4	Pencils	4	0.76	0.76			
		Stamps	5	1.00	1.00			
	5	Taxi fare	–	2.25		2.25		
		Biscuits	6	1.00				1.00
	6	String	7	0.65				0.65
				13.98	3.76	2.93	4.00	3.29
	6	Balance c/d		6.02				
20.00				20.00				
6.02	8	Balance b/d						
13.98	8	Imprest refund						

At the end of the week the main cashier will refund to the petty cashier the sum of £13.98 (the total disbursements) so that the balance of the imprest for the commencement of the second week is made up to the original sum of £20.00.

As far as possible, vouchers and receipts for payment of these small expenses should be asked for, so that they can be verified by the main cashier. It should be possible to obtain a receipt for most of the items.

If there is no receipt (and ideally even if there is) a petty cash slip should be completed. Blank pads of petty cash slips can be bought from business stationery suppliers. The slip should show clearly exactly what the petty cash reimbursement was for, and the amount, as till receipts are often uninformative and frequently fade after a few months. Some businesses will also want to have the petty cash slip authorized by a manager before it is refunded.

KEY POINTS

▶ The word 'imprest' means loan. The imprest amount is loaned to petty cash from the business.

▶ The analysis columns in use refer to regular routine expenses, often office expenses. A 'sundries' or 'miscellaneous' column takes care of casual and incidental expenses.

▶ Sometimes an extra column is used for the payment of small and occasional payments to a few creditors' accounts in the bought (suppliers) ledger.

▶ Where a full length exercise includes a petty cash book, remember to bring the petty cash balance on to the trial balance at the close of the trading period, and again on the assets side of the balance sheet.

▶ Remember, too, that all the *analysis expense totals* must be taken to the *debit* of nominal ledger accounts, and brought to the debit of the trial balance. This would not apply, though, to the settlement of small bought ledger accounts (see third bullet above).

TESTING YOURSELF

13.1 Willie Woodbine is the petty cashier of a small textile manufacturer. He is responsible for a weekly imprest of £30. During his absence on holiday, you are required to write up his petty cash book for the first week of March, using four analysis expense columns. Show the amount of the imprest refund at the end of the week.

March		£
1	Petty cash imprest in hand	30.00
	Bought stamps	3.00
2	Paid for travelling	3.20
	Milk for office	0.82
3	Newspaper	0.65
	Paper	2.50
4	Bought stamps	3.00
	Paid bus fare	1.28
	Parking	0.50
5	Wrapping paper and string	1.48

13.2 Rule up a petty cash book with five analysis columns and enter the routine expenses shown below. Show the imprest refund at the end of the first week in May.

May

1   Petty cash in hand £35

2   Bought stamps £4; paid £3.85 for stationery; and 75p for tea.

3   Paid cleaner £5; 52p excess postage due and £2.50 taxi fare for a sick employee.

4   Bought £4 stamps; refunded 36p bus fare to newly engaged office junior.

5   Window cleaner £2.35; paper £2.26; donation to police charity £2.

13.3 Explain what is meant by 'petty cash imprest'.

Rule up a petty cash book with four analysis columns and with an imprest of £50. Make up your own headings and expense details for two weeks, balance up, and show the imprest refund.

13.4 Why do we need a separate petty cash book?

13.5 What is the 'sundries' column used for?

14

Credit transactions and suppliers' accounts

In this chapter you will learn:
- *how to keep a purchases day book*
- *how the documents for purchases are used in accounting*
- *about trade and cash discounts*

All trading transactions, up to this stage, have been on a cash (or cheque) immediate payment basis. Modern business, however, is built upon credit: getting goods or services today, but delaying the payment and settlement for a few weeks or even a month or two. This means that permanent records must be kept of these credit transactions, and also of debtors and creditors.

A debtor (also known as a **receivable**) is a customer who has bought goods or obtained a service *on credit*, i.e. the goods or the services received have not been paid for yet. *Money is owed by a debtor.*

A creditor (also known as a **payable**) is the supplier of goods or services, who has not yet been paid. *Money is owing to a creditor.*

Types of ledger

Up to now we have assumed that all the accounts are in one ledger. However, in practice this quickly becomes unwieldy as soon as there are a number of customers and suppliers. Each of them needs their own separate account, so that we can show what we owe them, or what is owed by them to us. We could keep them in alphabetical order, but suppliers would get mixed up with customers, and accounts such as fixtures and fittings, or purchases, would be hard to find.

The two largest categories of accounts are those of suppliers and of customers. These are therefore normally separated out into two separate ledgers – the purchases ledger and the sales ledger. Total purchases then get posted to the purchases account, total sales to the sales account. The method of doing this (using day books) is explained below. The rest of the double entry accounts are kept in a ledger known as either the general ledger, or more often the nominal ledger.

Day books or journals

Subsidiary books called **day books** or **journals** are used for the purchases and sales of goods or services on credit. These books, referred to as **original records of entry**, are used as aids for posting the ledger accounts. Generally, after the extraction of essential information the books are stored away, to be produced for checking by the firm's auditor when required.

An illustration is now given of this new procedure as applied to the purchases side of the business and the trade creditors' accounts.

THE PURCHASES DAY BOOK

As explained in Chapter 2, the primary accounting document for merchandise bought on credit is the supplier's bill or invoice, handed over to the purchaser by the vendor at the time of the purchase or sent to the purchaser's address by post.

An invoice is shown first, the essential information then being extracted from it and posted to part of a folio of the purchases day book, or **bought journal** as it is sometimes called.

Samuel Smith South Parade Seatown	**INVOICE** DAVID BROWN **5 Eastgate Stafford**			No. 876 3 Feb. 20..	
Stock Ref.				£	£
Y 38	60 white cotton T-shirts size 16 @ £3.60			216	
	Less 25% trade discount			54	162
P 54	360 pair Nustyle socks @ £1			360	
	Less 20% trade discount			72	288
	E.&O.E.		Net		450

The ruling of the purchases day book is similar to that of the journal, the paper generally used by students for trial balances.

Purchases Day Book (36)

Feb. 3	Invoices from suppliers David Brown Stafford	BL 52			£ 450

The postings from another three invoices are now added to the **purchases day book** (PDB), to show the double entry procedure between the day book and the personal and nominal ledger accounts:

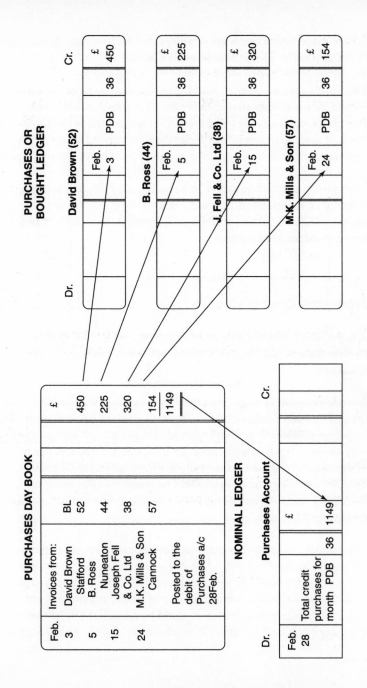

PURCHASES OR BOUGHT LEDGER

Dr.   Cr.

David Brown (52)

			£
Feb. 3	PDB	36	450

B. Ross (44)

			£
Feb. 5	PDB	36	225

J. Fell & Co. Ltd (38)

			£
Feb. 15	PDB	36	320

M.K. Mills & Son (57)

			£
Feb. 24	PDB	36	154

PURCHASES DAY BOOK

Feb.	Invoices from:	BL	£
3	David Brown Stafford	52	450
5	B. Ross Nuneaton	44	225
15	Joseph Fell & Co. Ltd	38	320
24	M.K. Mills & Son Cannock	57	154
			1149
	Posted to the debit of Purchases a/c 28Feb.		

NOMINAL LEDGER

Purchases Account

Dr.   Cr.

			£
Feb. 28	Total credit purchases for month PDB	36	1149

Where value added tax is applicable, separate columns are needed in the day book. This is explained in Chapter 17.

The purchases day book is used for goods bought on credit for re-sale, or for re-sale after conversion. *Permanent assets acquired such as fittings and office equipment are not posted in the day book*: if these are bought on credit, they are debited direct to the asset account and credited to the personal account of the supplier.

Cash purchases (which may also be paid for by cheque) are taken direct from the credit of the cash book to the debit of a separate nominal ledger account, and the cash memo details are filed in a separate folder for easy access by the auditor.

Trade and cash discounts

A topic which often comes up in examinations (probably more so now than in real life) is the difference between trade and cash discounts.

A **trade discount** is most often used by a supplier who sells both to the public and to other traders. Because other traders will normally be bigger customers, they are offered cheaper prices. The supplier may simply keep two separate price lists, but it is often easier to offer traders a flat percentage off the price charged to customers. This is known as a trade discount. Questions will therefore say something like 'Mr Jones sells £1000 of goods less a trade discount of 20% to Mr Smith'.

The point to remember is that **trade discounts never get entered separately in the day books.** Not in the purchases day book, nor in the sales day book that you will learn about in the next chapter. You calculate the discount and deduct it before making the entry. The entry in the example above would be £800 in both Mr Jones' sales day book and Mr Smith's purchases day book. This is logical, because Mr Smith has never agreed to pay more than £800, nor was Mr Jones expecting to receive more from a trade customer.

Cash discount, on the other hand, is a discount for prompt payment. Continuing the example, Mr Jones might offer a 2% cash discount to Mr Smith if he pays within 14 days. If Mr Smith does so, he only needs to pay £784. When he pays this, the bookkeeping entry for Mr Jones will be to debit cash and credit Mr Smith's account. This, however, still leaves £16 (the cash discount) showing as a balance owing, when in reality Mr Smith owes nothing. To clear the £16, a further entry is made: credit Mr Smith and debit a new account, 'Discounts allowed'. To test that you understand this, stop now and write out the three accounts in Mr Jones' books – Mr Smith's customer account, the cash account and 'Discounts allowed'. You should end with a balance of £784 cash, £16 discount allowed and nothing owed by Mr Smith.

In Mr Smith's accounts, the original entry was to debit purchases and credit Mr Jones' account with £800, because that is the price he has agreed to pay. When the £784 was paid, taking advantage of the offer of 2% cash discount, the entry was to debit Mr Jones £784, credit cash £784. Again this leaves £16 to clear, so a further entry is made – debit Mr Jones £16 and credit 'Discounts received' £16. Again, stop and draft out the accounts to check that you have understood this.

Rather than create separate ledger accounts, cash discounts are often dealt with by adding a third column to the cash book, so that it now reads 'discount', 'cash' and 'bank'. In the above example when the bill was paid the entry would have been to debit the personal account with the full £800 with the contra shown as 'cash book', and then £16 and £784 entered in the discount and bank columns respectively. This is known as a **three-column cash book**. When it is ruled off at the end of the period, the debit and credit totals for cash and bank will be offset as usual to give a figure to take to the trial balance. However, the totals of each of the discount columns are taken to the trial balance separately – so far as the double entry system is concerned they are still separate accounts, even though they have been recorded in the cash book for convenience.

Because separate accounts are created for cash discounts allowed and received (either as ledger accounts or in a three-column cash book), **cash discount will always appear as a separate entry in the profit & loss account.**

To operate a cash discount, a **three-column cash book** is required, showing discount as well as cash and bank. Cash discount is entered in the discount column of the cash book on the left of the payment to which it relates, and the combined total of the *money paid plus the cash discount* allowed or received is posted to the personal account of the credit customer or the supplier of the merchandise.

Insight

A general mistake that is often made is to treat cash discounts as an adjustment to selling or purchasing price, rather than as a separate profit & loss item; and also to include trade discount in discounts received/allowed rather than adjusting the price. If you understand the principles set out above, it should be clear when a discount is trade or cash, and therefore how it should be treated. If you really cannot work out whether a discount is cash or trade, it is worth noting that cash discounts are normally a very small percentage – typically 5% or less. Trade discounts are normally 20% or more.

Purchases account

The same purchases account is used for both cash and credit purchases, but now there is a slight change of procedure for credit purchases. The individual debits (invoice totals) are not posted to the nominal ledger at the time they are entered in the day book, but are allowed to accumulate day by day and are *posted in one total for the month* to the purchases account at the end of each month.

The individual supplier's (creditor's) account, however, is posted up immediately, direct from the day book, so that at any time the amount owing to any creditor may quickly be ascertained.

Creditors' accounts

The balances outstanding (unpaid) on all suppliers' accounts are listed at the end of the trading period. The total for 'trade creditors' is taken to the credit side of the trial balance, its final destination being under the heading of current liabilities on the balance sheet.

Payments made to creditors in settlement of their accounts *does not affect the purchases total* on the debit of purchases account.

Cash or credit?

Students sometimes experience a little difficulty in deciding whether a transaction is for cash or on credit. These rules will help.

▶ *Unless instructed to the contrary, assume that all transactions for the purchase or the sale of goods are on credit when a personal name or the name of a firm is mentioned in the question. There should be no need to mention the name of the buyer or vendor in the cash of a cash purchase or a cash sale as money is simply exchanged for the goods handed over.*
▶ *Assets and office equipment are often bought on credit, and again this should be assumed when the name of the supplier is mentioned, but remember that the cost of the asset should be credited direct to the supplier's account and the debit taken direct to the asset account.*
▶ *Cash purchases, cash sales and all transactions where the word 'paid' is mentioned are obviously cash transactions.*

The buyer's records

Accounting systems and rulings vary a good deal, but the general principles involved in the purchase and sale of goods remains basically the same. On the purchase side, illustrated below, start with the official order placed with the supplier and follow through the invoicing process and the various aspects of documentation. If you cannot remember what happens when one business orders goods from another, refer back to Chapter 2.

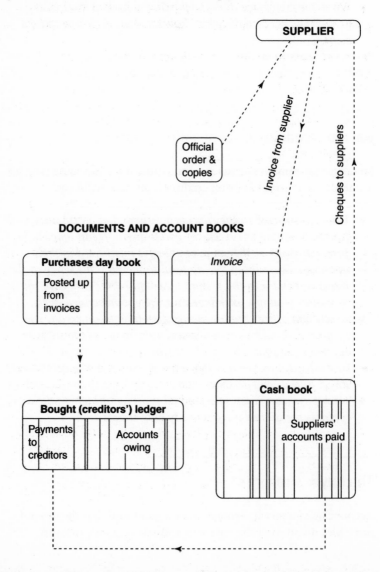

PAUSE FOR THOUGHT

▶ *What is a book of prime or original entry? Does the cash book fall into this category?*

▶ *What kind of purchases are entered up in the bought journal?*

▶ *Would you expect to find the purchase of a new car posted to the bought journal? If not, how would you deal with this acquisition?*

▶ *What is another name for the bought journal?*

KEY POINTS

▶ Money is owing to trade creditors. Their ledger accounts kept in the bought ledger, for merchandise supplied on credit, have not yet been paid.

▶ The purchases or bought day book is a book of prime entry, used for recording invoices of credit purchases (goods to be re-sold or material bought with a view to reconditioning it for sale). Other books of prime entry are the cash book, the sales day book and the returns books.

▶ Suppliers' accounts, kept in the bought ledger, have credit balances. The individual credits have been posted to these accounts day by day from the bought day book. The day book totals are carried forward until the end of the month, and then the monthly total is taken to the debit side of the purchases account in the nominal ledger, thus completing the double entry.

▶ Trade discount is an allowance made from the vendor's price list or catalogue. The deduction is made from the invoice and the net amount due to the creditor is posted net to the credit of the supplier's account in the bought ledger.

▶ Payments made to the creditors do not affect the day book or the total taken to purchases account. These payments concern only the cash book and the personal accounts of creditors.

TESTING YOURSELF

14.1 Enter up the bought day book of Maree Lyritis from the details below, and post up the double entry to the bought ledger.

July
4   Bought £800 of goods from Adrian Shaw less 20% trade discount.
9   Received delivery of £285 goods from Samuel Swift. This included £25 transport costs.
18 Bought another £600 of goods from Adrian Shaw less usual T.D.
24 Ordered by telephone £450 of goods from W.H.T. Suppliers Ltd. Forwarded and invoiced the following day.
30 Received a debit note for £180 from Samuel Swift for goods delivered and checked the previous day.

Assuming Miss Lyritis settled the two earlier accounts of Shaw and Swift on 20 July, show all ledger accounts in full detail and balance up on 31 July.

14.2 Garry Hall commenced trading with £8000 in the bank and a stock of goods valued at £2000 on 1 April. His transactions during April are listed below.

April		£
2	Cashed cheque for office	500
	Bought goods from Tom Wynne £600 (gross) arranging for trade discount of 25% on all purchases	
4	Cash sales	1263
	Paid for stationery (cash)	84
	Bought stamps	40
6	Bought goods from Fred Maples	366
	Cash sales	1568
	Paid cash into bank	2000
	Bought goods from Wynne (gross)	800
12	Paid Maples on account	200
	Paid insurance	155
	Sent cheque for advertising	122

18	Cash sales	1627
	Paid into bank	1000
	Cash purchases (cheques)	182
	Paid Wynne's April account	1050
30	Cheque for month's wages	3200
	Cash for 'self'	600

Post all transactions through original records to the ledger accounts. Balance all accounts and take a trial balance at 30 April.

14.3 Explain what is meant by a 'payable' and a 'receivable'.

14.4 Why are customer and supplier accounts separated out into two separate ledgers?

14.5 What are the original records of entry?

14.6 What is the difference between a trade discount and a cash discount?

15

The accounts of credit customers

In this chapter you will learn:
- *how to keep a sales day book*
- *how sales documents are used in accounting*

When goods are sold on credit, the ownership of the goods passes immediately to the buyer who, at the time of the purchase or perhaps a few days later, receives an invoice from the vendor showing the full details of the sale, with trade discount deducted if this is applicable, and the net amount to be paid in due course.

> **Insight**
> Remember the key document relating to credit sales made by the business is the invoice. Also that the credit note, which looks like an invoice, but states that the amount owing to you by the customer is to be reduced by the amount shown.

Sales day book

In the books of the vendor, credit sales follow a similar procedure to that already outlined on the purchase side in the previous chapter, another book of original entry being used called the sales day book (SDB).

Before computerization, large businesses made several copies of each invoice sent to their credit customers, one copy being taken by the accounts department and used for making up the sales day book.

The essential details (date, name of customer, and net amount) were taken from the copy of the invoice sent to the credit customer and entered in the sales day book, trade discount being deducted (if allowable) to show the net charge to the customer. Only this *net amount* was posted to the debit of the customer's sales ledger account.

The daily sales total was carried forward and at the end of the month the total gross sales were credited to sales account in the nominal ledger. The illustration shows the posting procedure between the day book and the personal and nominal ledgers.

In any system that uses a sales day book, it is used only for *credit sales*. Cash sales and miscellaneous receipts are credited to a nominal ledger account unless in connection with the sale of assets (money received from the sale of old fittings would be debited to cash and credited to fittings account).

Remember that payments in settlement of personal accounts *do not affect sales*. Payments by debtors simply involve the cash book and the personal accounts of the credit customers.

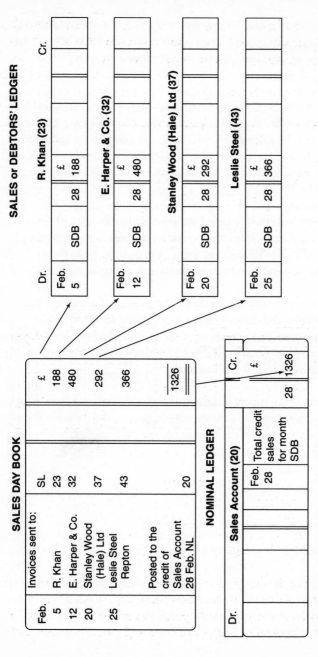

Illustration   The posting procedure between day book and ledgers

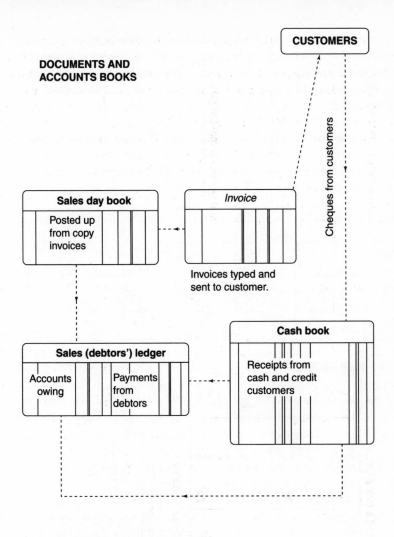

DOCUMENTS AND ACCOUNTS BOOKS

CUSTOMERS

Cheques from customers

Sales day book
Posted up from copy invoices

Invoice

Invoices typed and sent to customer.

Cash book
Receipts from cash and credit customers

Sales (debtors') ledger	
Accounts owing	Payments from debtors

Columnar bookkeeping

Day books may be designed and adapted to suit the needs and purpose of the business. If there are several departments the original books of entry for credit purchases and credit sales can be ruled to show a detailed analysis of costs and the revenue for each department. Separate trading accounts can then be made up, with stocks attributable to each department. Comparisons may then be made between the departments, or between the different product lines and the non-profit making lines discarded, or economies put into effect to improve profitability.

A break-down on the purchase side of a business with three separate departments might take the following form, in so far as the bought journal is concerned (VAT is ignored for ease of comprehension).

Date	Supplier	Inv. No.	BL folio	Total	Dept. X	Dept. Y	Dept. Z
June				£	£	£	£
1	S. Adams	36	24	120	55	65	–
3	L.H. Dale	45	16	14			14
	Guy Lord	46	176	240	150	40	50
6	Jay & Lee	47	82	72		48	24

KEY POINTS

▶ Money is owing by debtors. The listed total for trade debtors is taken to the debit of the trial balance and shown under the heading of 'current assets' on the balance sheet.
▶ Brief details are taken from copies of the invoices sent to customers, to make up the sales day book.
▶ Neither the sales day book nor the sales account are affected by the individual payments from debtors, the accounting entries involved simply concerning only the cash book and the customer's personal ledger account in the sales or debtors' ledger.

TESTING YOURSELF

15.1 Felicity Jones commenced business on 1 October with £5000 in the bank and a stock of goods valued at £2000. These are her recorded transactions for the month of October.

Oct.		£
1	Cashed cheque for the office	500
	Ordered £400 of goods from	
	Sarah Gosling subject to 20% trade discount.	
	The merchandise was delivered the next day.	
3	General expenses paid in cash	60
	Paid for some advertising by cheque	155
	Cash sales	724
9	Cash purchases	91
	Cheque drawn for 'self'	300
12	Cash sales	872
	Paid cash into bank	1000
	Sold goods (on credit) to Tom Lister	258
15	Bought £600 of goods from Roy Morris less 10%,	
	and sold £450 of goods to Ted Batten making him	
	an allowance also of 10%	
18	Cash sales	667
	Paid Sarah Gosling's account	
20	Received cheque from Tom Lister	258
	Sold £300 of goods to Ted Batten, invoiced gross less 10%	
25	Drew cheque to pay salary of part-time receptionist	900
	Withdrew cash for private use	500
30	Cash sales	586
31	Paid all surplus cash into bank except for a cash	
	float of £200	

You are required to post up all the original records, up balance all ledger accounts, and take a trial balance on 31 October.

15.2 What is a sales day book used for?

15.3 Where are cash sales posted to?

16

Purchases and sales returns

In this chapter you will learn:
- *how to account for returned goods*

In Chapter 2 we looked at what happened when a business had to return purchases, and saw the documents that are created – credit notes and debit notes. Reread this section of Chapter 2 if you cannot remember it. We are now going to see how to enter these in the accounting records of the business.

Accounting for returns

The accounting procedure is not complicated. It simply involves putting the whole or part of the original transaction in reverse. When goods are returned to your supplier, your *total purchases are reduced*, and, if bought on credit, you do *not owe your creditor so much*. The original transaction or part of it is reversed by debiting your supplier's account and crediting purchases with the value of the goods returned.

Similarly on the sales side. When the credit customer returns some goods to the supplier, the figure for the *recorded sales is reduced* (in the books of the vendor) and the *debtor does not owe so much*. The bookkeeping procedure is a debit to sales account and a credit to the personal account of the customer (debtor) for the value of the goods returned.

One important thing to remember about the return of goods is that any **trade discount** deducted on the original purchase or sale must also apply and be *deducted from the gross value* of the goods which are returned.

In practice, when a trader's returns of goods sold are few and infrequent, brief details of the returns are often recorded at the back of the ordinary purchases and sales day books.

In the larger firms, and particularly where returns are more numerous, separate returns day books are brought into operation. These are known as the **returns outwards book (purchase returns)** and the **returns inwards book (sales returns)**.

The ultimate destination of the returns is to the credit of the purchases account or the debit of the sales account, but in the meantime the trial balance may be made up. Assuming that the returns have already been taken in reduction of purchases and sales account, the *net purchases* and the *net sales* figures are debited and credited to the trial balance, but it would not be incorrect to show the gross figures for the purchases and the sales together with separate amounts for the returns, i.e. a debit for returns inwards (sales returns) and a credit for returns outwards (purchase returns).

ILLUSTRATION

Let us assume that two of the white cotton T-shirts invoiced by David Brown on 3 February (Chapter 14) were torn in transit, and that a claim is made by the buyer Samuel Smith against the full charge shown on the original invoice no. 876 dated 3 February.

David Brown would investigate his customer's complaint and examine the faulty merchandise. He acknowledges the claim by sending a credit note, say on 8 February, to Samuel Smith, for the value of the two T-shirts already charged, i.e. £7.20 *less trade discount* at 25% which was deducted from the original invoice.

In Samuel Smith's books (the purchaser's books) the posting from the returns outwards book to the creditor's bought ledger account is shown, which reduces the balance due to this supplier by £5.40. The purchases account total for the month is also reduced by the same amount by a credit for these returns which is posted from the returns outwards book at the end of February.

In the illustration shown, the returns outwards book total for February has been taken direct to the credit of purchases account at the end of the month. In some instances an account might be opened for these purchase returns called purchase returns account or returns outwards account and the credit total for the month posted to this intermediary account, and then a **transfer** made of the returns to the credit of purchases account when the trading account is made up. In any event, the returns will ultimately be deducted from the purchases total in trading account, so that the figure for *purchases will be shown net*.

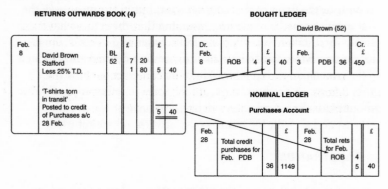

A similar procedure is adopted with the sales returns. The net value of the goods returned is taken straight from the returns inwards book to the credit of the customer's personal account in the sales (debtors') ledger, reducing his indebtedness. The total of the returns inwards book for the month will be taken to the debit of sales account (or an intermediary returns inwards account if one is in use), and the turnover figure on trading account will be shown *net*, i.e. total sales less returns inwards.

Where VAT is applicable, again there will be the need for additional columns in the returns books (see the next chapter).

TESTING YOURSELF

16.1 Write up the account of Julian Morton in the books of The Midland Agency from this information for the month of March.

March		£
1	Balance brought forward	844
3	Credit sales to Morton	562
	He paid his February account	
12	Further sales to Morton	363
	Allowance for returns	25
22	Bought some special fittings and equipment from Morton	226
25	Returned some damaged parts bought on 22 March.	
	Morton sent credit note to adjust	48
31	Balanced the account	

16.2 Explain what credit notes and debit notes are.

16.3 What are the returns outwards book and the returns inwards book used for?

17

VAT and PAYE

In this chapter you will learn:
- *how to account for PAYE*
- *how to account for VAT*
- *what the balance on the Customs & Excise account means*

Introduction

Any business will have to deal with tax, but initially the only contact with the tax authorities may be for the business proprietor to pay **income tax** on the profit made. However, as a business expands there are two other taxes it is most likely to come into contact with, and where it will act as an unofficial tax collector for the Government. These are **Value Added Tax** (VAT) and **Pay As You Earn** (PAYE), the system for deducting tax and national insurance from the wages paid to employees.

At this level, no detailed knowledge of either is required, but you do need to know how to deal with the accounting entries that they may generate. It helps to understand those if you learn a little about the taxes involved.

Value Added Tax

VAT is a tax on goods and services. As its name suggests, it is intended to be a tax on the value added at each stage, until the product or service is bought by the consumer.

EXAMPLE

Peter Smith buys sheet steel from John Brown for £100 and forges it into 100 spoilers for cars. He sells the 100 spoilers to Iqbal Aziz for £250. The added value is £150 (£250–£100).

Iqbal Aziz spray paints the spoilers so that they will match the cars. He sells them to David Adams for £450. The added value is £200 (£450–£250).

David Adams fits the spoilers to customers' cars for £10 each, plus tax. If he can sell and fit all 100 spoilers he makes £1000, and the added value is £550 (£1000–£450).

For the sake of the example, assume that VAT is 10%. When Peter Smith buys the steel for £100, the vendor will add 10%, so he will actually pay £110. John Brown has to hand over the £10 VAT to HM Revenue and Customs, who are responsible for administering the VAT system.

When Peter Smith sells the spoilers he will also add VAT to the **full** sale price, a tax charge of £25. When he works out what he owes HM Revenue and Customs, he is allowed to deduct the £10 VAT he paid out from the £25 collected, and pay over the balance, £15. This, it should be noted, is the VAT at 10% on the *value he has added* of £150.

Similarly Iqbal Aziz adds £45 to the cost of the spray painted spoilers he sells to David Adams. Mr Aziz pays HM Revenue and Customs £20 – the £45 VAT he has collected *less* the £25 he has paid out. Again this is 10% of the value he added, £200.

Finally David Adams adds £1 to the £10 cost of each of the spoilers he sells and fits. He collects £100 in VAT, deducts the £45 he paid, and hands over £55 to HM Revenue and Customs.

In total HM Revenue and Customs have collected 10% of the final sale price of the spoilers – £100 in all, but they have collected it from each trader in proportion to the value he added.

Trader	Value added	Tax due	
John Brown	£100	£10 – 0 =	£10
Peter Smith	£150	£25 – 10 =	£15
Iqbal Aziz	£200	£45 – 25 =	£20
David Adams	£550	£100 – 45 =	£55
Total	£1000		£100

A business does not have to register for VAT unless its taxable supplies in the previous twelve months exceed £68 000, although it may register voluntarily with a turnover below this figure. This figure, and the VAT rate and fraction given in this chapter, are as at April 2009. A business which is not registered for VAT will enter all its costs inclusive of any VAT charged, it will not separate out the VAT. A business which is registered has to have separate analysis columns for the VAT, as will be explained below.

Dependent on the particular supply of goods or services, a supply may be **exempt** from VAT, **zero-rated**, charged at the **standard rate** which is currently 17.5% or lower rate which is 5% in a few cases. Although there are important differences for tax purposes between exempt and zero-rated supplies, the consequence for bookkeeping is the same – no VAT is included. All other supplies which the business pays for (known as the **inputs**) will include a sum for VAT, and all supplies it makes (known as the **outputs**) will have to have VAT added to them. If the invoice for an input only gives a total cost including VAT, the amount of VAT can be found by using the current VAT fraction. At present it is $\frac{7}{47}$, so if an item costs £94 including VAT at 17.5%, the VAT is $\frac{7}{47}$ × £94 = £14, and the price before VAT is £80.

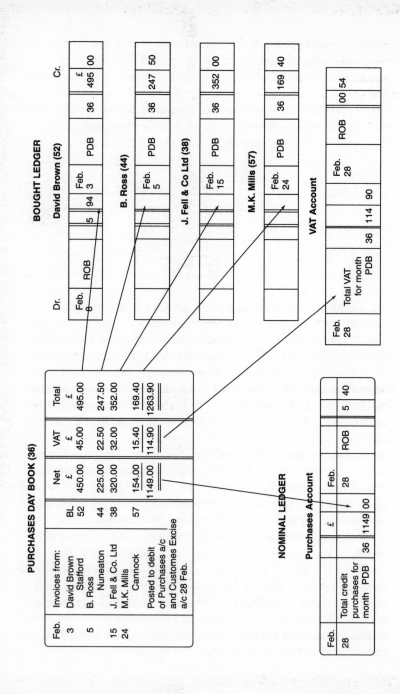

PURCHASES DAY BOOK (36)

Feb.	Invoices from:	BL	Net £	VAT £	Total £
3	David Brown Stafford	52	450.00	45.00	495.00
5	B. Ross Nuneaton	44	225.00	22.50	247.50
15	J. Fell & Co. Ltd	38	320.00	32.00	352.00
24	M.K. Mills Cannock	57	154.00	15.40	169.40
			1149.00	114.90	1263.90
	Posted to debit of Purchases a/c and Customes Excise a/c 28 Feb.				

BOUGHT LEDGER

David Brown (52)

Dr.

| Feb. 8 | ROB | | 5 | 94 | Feb. 3 | PDB | 36 | 495 | 00 | Cr. £ |

B. Ross (44)

| | | | | | Feb. 5 | PDB | 36 | 247 | 50 |

J. Fell & Co Ltd (38)

| | | | | | Feb. 15 | PDB | 36 | 352 | 00 |

M.K. Mills (57)

| | | | | | Feb. 24 | PDB | 36 | 169 | 40 |

VAT Account

| Feb. 28 | Total VAT for month PDB | 36 | 114 | 90 | Feb. 28 | ROB | | 00 | 54 |

NOMINAL LEDGER

Purchases Account

| Feb. 28 | Total credit purchases for month PDB | 36 | 1149 | 00 | Feb. 28 | ROB | | 5 | 40 |

122

In order to record the VAT, the day books of a VAT-registered trader will have extra columns to account for VAT. On the purchases side, there will now be three columns – **Net, VAT** and **Total**. The amount in the Total column for each transaction, the total of the net cost plus the VAT, is the amount which is posted to the credit side of the supplier's account.

At the end of the month or whenever the books are posted, the separate totals for net cost and VAT are debited respectively to the purchases account and to the VAT account in the nominal ledger.

Similarly on the sales side there are again three columns – Net, VAT and Total. The amount in the Total column for each transaction is posted to the debit side of the sales ledger for that customer, and in due course the totals of the net and VAT columns are credited to the sales account and to the same VAT account in the nominal ledger as was used for the purchases. A similar process is followed for returns.

Returning to the illustration of the purchases day book procedure set out in Chapter 14, the transactions are now shown with VAT included. To make the calculations simpler the VAT rate is assumed to be 10% – in an examination you will be told what rate of VAT to use.

It can be seen that the figure debited to the purchases account remains the same: £1149 less the £5.40 of returns, giving a total of £1143.60. However, the amounts owed to Brown, Ross, Fell Ltd and Mills have increased by a total of £114.36, so in due course they will have been paid £1257.96. The extra £114.36 is the £114.90 debited to the VAT account less the 54 pence credited on the returns.

The sales side is dealt with in a similar way – look again at Chapter 15. When the additional columns have been added as explained above, and VAT at 10% (say), £132.60 would be added *in total* to the debtors' accounts. The amount credited to sales in total would still be £1326, and the additional £132.60 would be credited to the same VAT account. It is the net balance on the account that is due to HM Revenue and Customs (or repayable if it is a debit balance), which in this case is £18.24. So the VAT account at the end of the month would look like this:

Dr.			£	Feb.	VAT Account		Cr. £
28	VAT input total for February PDB	36	114.90	28 28	Returns outward book VAT output total		00.54
28	Balance due to H.M. Revenue and Customs c/d		18.24		for February		132.60
			133.14				133.14
Mar.				Mar.			
				1	Balance b/d		23.10

The normal pattern is that a business produces quarterly **VAT returns**, submitting them together with any payment on or before the end of the month following the end of the quarter. So, for example, a business on a January/April/July/October quarterly cycle would have to deliver a return for the three months to January by the end of February, for the three months to April by the end of May, and so on. It follows that the VAT account will not normally ever drop to zero, since by the time the VAT due for the three months to January is paid, transactions for February will have been entered.

Insight

Its important to note that, in accounting terms, this money does not 'belong' to the business. All that the business is doing is acting as an unpaid tax collector: collecting VAT on behalf of HM Revenue and Customs on the goods or services it sells, and offsetting the VAT it has to pay on its purchases. The net balance is handed over to HM Revenue and Customs on a periodic basis.

HM Revenue and Customs (VAT) Account
4 months ended February

VAT input totals from PDB			VAT outputs totals from SDB		
30 Nov.	Nov. total	3100	30 Nov.	Nov. total	4800
31 Dec.	Dec. total	2400	31 Dec.	Dec. total	3000
31 Jan.	Jan. total	2800	31 Jan.	Jan. total	3700
31 Jan.	Bal. c/d	3200			
		11500			11500
			1 Feb.	Bal. b/d	3200
28 Feb.	Feb. total	2200	28 Feb.	Feb. total	3300
28 Feb.	Bank	3200			

PAYE

Pay as you earn, or PAYE, is the system under which the income tax and national insurance contributions due from *employees* is deducted from their pay by their employer who then pays it on to HM Revenue and Customs.

The details of how the tax and national insurance due are calculated is beyond the scope of this book, and is not covered in accounting or bookkeeping examinations. You only need to be aware of the basic terms involved, and how the transactions are entered in the books.

An employee will be entitled to a certain amount of **gross pay**, which may vary with overtime, piecework, holidays etc. Say for example that for a given week an employee's gross pay is £200. Using a code number issued by HM Revenue and Customs and a set of tax tables, the employer will calculate the income tax due on this pay under PAYE. Say for example this comes to £30. The employer then uses another set of tables to calculate the national insurance contributions (NIC) due – this might be £15. The total of these two, £45 in our example, is often referred to as the **deductions from gross pay,** or simply as **deductions**. This amount is not paid out to the employee, but is kept back by the employer and paid over on a monthly basis to the HM Revenue and Customs. The employee actually receives £155 in our example (£200–45) which is called the **net pay**.

Again, in order to understand the accounting it is important to understand what is happening. The employee is actually entitled to £200 in pay. However, this employee also has to pay £45 to HM Revenue and Customs. To make sure that they get their money, the PAYE system requires the employer to hold this amount back and only give the employee the net pay, with the balance being paid over to HM Revenue and Customs in settlement of the *employee's* tax liabilities.

Assume that there are ten employees all paid the same as the one in the above example. The total gross pay for the business is therefore £2000, and this is the amount debited to the wages account. The cash actually drawn out to make up the ten employees' wage packets will be the total net pay of £1550, so this is the amount credited in the bank account. The remaining £450 is credited to the PAYE/NIC account, showing HM Revenue and Customs as a creditor for this amount until it is paid over at the end of the month.

There are two further complications.

▶ *As well as the NIC that is deducted from the employee's pay, the* **employer** *also has to make NIC payments based on the wages paid to the employees. The accounting entries for these are simply to debit the wages account again and credit the PAYE/NIC account.*
▶ *The second point to note is that it is only the tax and national insurance in respect of employees that are dealt with through the PAYE system. The income tax and national insurance paid*

by a sole trader or a partnership based on the trading profit is not an expense of the business at all – it should be treated as drawings. However, if the business is run as a limited company, the main shareholders may well be directors. Any salary that they are paid is subject to PAYE in the normal way.

TESTING YOURSELF

17.1 Explain the terms VAT and PAYE.

17.2 At what level of turnover must a business register for VAT?

17.3 In terms of VAT what are inputs and outputs?

17.4 When should a VAT return be submitted?

17.5 What is the difference between gross pay and net pay?

18

Classification of ledger accounts

In this chapter you will learn:
- *how to classify ledger accounts as personal or impersonal*
- *how to classify impersonal accounts as real or nominal*
- *what this means when drawing up a balance sheet*

There are two main divisions of ledger accounts:

▶ **personal accounts** *of people, firms and companies relating in the main to the credit accounts of suppliers and customers.*
▶ **impersonal accounts,** *subdivided into:*
  ▷ *real and property accounts, comprising 'fixed assets' such as premises, machinery, cars and vans, furniture and fittings, and 'current assets' of stock on hand, money at the bank and in the office*
  ▷ *nominal accounts, relating to the firm's sales revenue and the expenditure incurred in creating that revenue. These accounts include profits and gains (sales, commissions earned, rents and discounts received) and also losses and expenses such as wages and salaries, advertising, rent and rates payable, repairs, lighting and heating, insurance, and miscellaneous expenses.*

General rules for posting

	Personal accounts	Real/property accounts	Nominal accounts
Debit:	Who receives	Asset value coming in	Losses and expenses
Credit:	Who gives or pays	Asset value going out	Profits and gains

It is customary for all personal ledger accounts to be balanced at the end of the month, the *new balance being brought down* showing the exact amount owing to each creditor or due by each debtor. When, however, an account shows a single posting, it need not be balanced, as the *amount shown is the balance*.

In large firms credit customers' accounts are often so numerous that further sub-division of the sales ledger is necessary. It could be divided into three or more separate parts, alphabetically or geographically, for instance.

Common forms of ledger accounts

Dr.   **Andrew White** (debtor's *personal* account)   Cr.

June			£	June			£
1	Balance b/f		10.50	3	Cheque	CB	10.50
15	Sales	SDB	18.20	17	Returns	SRB	2.20
				30	Balance c/d		16 00
			28.70				28.70
July							
1	Balance b/d		16 00				

John Norton (creditor's *personal* account)

June			£	June			£
2	Cheque	CB	22.90	1	Balance b/f		22.90
6	Returns	PRB	3.60	5	Purchases	PDB	65.60
20	Balance c/d		96.20	18	Purchases	PDB	34.20
			122.70				122.70
				July			
				1	Balance b/d		96.20

Furniture & Fittings (a *real* account)

June			£				
1	Balance b/f		200.00				
8	Cheque (new cabinet)	CB	60.00				

Cars and vans (a *real* account)

June			£	June			£
1	Balance b/f		600.00	10	Sale of old van	CB	550.00
20	Cheque for new van	CB	2800.00	30	Loss on sale (P&L a/c)		50.00
				30	Balance c/d		2800.00
			3400.00				3400.00
July							
1	Balance b/d		2800.00				

Rates (a *nominal* expense account)

June			£			
24	Cheque for ½ yr's Rates	CB	320.00			

Insurance (*nominal* expense account)

June			£			
5	Fire & General	CB	88 00			
29	Employers' Liab.	CB	54 00			

Commissions Received (*nominal gains account*)

			June			£
			28	Cheque from WB Agency	CB	40 00

The destination of these ledger accounts from the trial balance to the trading and profit & loss account and the balance sheet is now shown:

	Trial Balance		T & P & L a/c		Balance Sheet	
	Debit	Credit	Debit	Credit	Assets	Liabs.
A. White, debtor	16.00				16.00	
J. Norton, creditor		96.20				96.20
Furniture/ fittings	260.00				260.00	
Motor vans	2800.00				2800.00	
Rates account	320.00		320.00			
Insurance account	142.00		142.00			

Loss on sale (motor van)	50.00		50.00			
Commissions rec'd		40.00		40.00		

The loss (and expense) on the sale of the old van has been taken direct from the motor vans (ledger) account to the debit of profit and loss account. After the closure of all the nominal (expenses and gains) accounts by transfer of their balances to the debit or to the credit of the revenue accounts, the only accounts left with 'open' balances are the asset, capital and liability accounts. These open accounts are carried forward to the next trading period, and are shown on the skeleton balance sheet below. The balance sheet cannot be completed as we only have a few of the accounts.

Assets employed		
Fixed assets	£	£
Van account	2800.00	
Furniture/fittings	260.00	3060.00
Current assets		
Stock	XXX	
Trade debtors	16.00	
Bank and cash	XXX	XXX
		XXX
Financed by		
Capital at 1 June	XXX	
Add net profit	XXX	
Less drawings	———	
	XXX	
Current liabilities		
Trade creditors	96.20	
Expense creditors	XXX	XXX
		XXX

Note that at this stage the whole of the profits and gains, losses and expenses of the business have been absorbed into the revenue accounts, ending with the net profit or loss on trading to be transferred to the proprietor's capital account.

PAUSE FOR THOUGHT

It is important to distinguish from a payment made (or a liability incurred) for a permanent fixed type of asset, and a revenue expense payment (or expense liability incurred). Later, in Chapter 22, this is explained in further detail.

Sometimes students are confused because *both assets and expenses have debit balances*. The main thing to bear in mind is that there is a credit entry, either in the cash book or on a creditor's account for *both asset and expense transactions*, but the ultimate destinations are quite different, as illustrated in this chapter.

Insight

Both asset and expense accounts have debit balances or totals. They are posted to quite different financial statements. The asset is taken to the balance sheet, whereas the expense is taken to the trading profit or loss account.

Classify the following under the headings of real, personal and nominal:

- *machinery*
- *bank overdraft*
- *capital*
- *new lease*
- *Judi Bray (customer)*

- *wages*
- *insurance (paid)*
- *rates in advance*
- *Delia Harlow (supplier)*

Which of these ledger accounts would have credit balances?

TESTING YOURSELF

18.1 List the following under account headings of real, personal or nominal:

Cash sales £54; van repairs £16; rates £200; commission earned £28; insurance paid £35; garage extension £850; new desk £80; new printer £180; sale of old printer £25; V.J. Pauli owes £45 for goods bought three months ago.

18.2 Post up the account of Howard Tarver, a wholesaler, from these details.

July
1	You owe Tarver £122.44 for goods bought in June.
3	Further purchases £60 less T.D. of 25%
5	Returns to Tarver £16 gross
	Paid his June account
	Bought further goods £72 less 25%
11	Sales to Tarver £25.50 less 20% T.D.
15	Tarver returned £5 gross goods
24	Further supplies from Tarver £36 less 25%. He also made an allowance of £2.60 for packing case returned
31	Balanced the account

18.3 Post up the account of Trevor Tongue in the books of Miller Bros and bring down the balance at 30 June.

June
1	Tongue owes Miller Bros £462 for supplies in May
3	Tongue buys further goods from Miller Bros to value of £800 less T.D. of 20%
5	He returns £100 gross to supplier, and settles May account
12	Further sales to Tongue £355 less usual discount
22	Cheque sent by Tongue for £500 'on account'. Allowance of £35 made for containers returned

25	Tongue's cheque returned by Bank to Miller Bros with remarks 'refer to drawer'
26	Tongue called on his creditor, handed in £400 in cash, tore up his cheque and promised balance early in July

Revision exercise 2

1 David Shaw commenced business on 1 July with a capital of £30,000 made up as follows.

	£		£
Cash in hand	300	Bank balance	23 450
Fixtures/fittings	1750	Van valued at	7000

An amount of £2500 was owing at 1 July to Roy Fenton for the van, recently acquired by the business.

After posting the above details to the cash book and the general ledger and the personal ledger account of Roy Fenton, you are required to record the transactions listed below for the month of July.

On completion of the postings, balance up all personal and nominal ledger accounts and take a trial balance at 31 July.

July		£	
1	Bought some shelving	1000	
	and a second-hand cupboard	200	
	Stamps	40	
3	Supplies bought for stock from		
	Bryan Benson, subject to a trade		
	discount of 20%	1200	gross
5	Cash sales	453	
	Sold goods to Esther Allen (allowing her		
	T.D. of £100)	700	gross
9	Cash sales	866	
	Paid cash into bank	800	
	Paid insurance (fire and theft)	125	
	Drew cheque for self	300	

12	Paid various advertising expenses	155	
	Bought further goods from Benson,		
	being allowed 20%	1500	gross
14	Returned goods to Benson	150	gross
	Cash sales	744	
	Sale at special price to Dinah Symonds	1040	
	Made urgent delivery to Owen Moore	388	
18	Cash purchases	120	
	Drawings for self (cheque)	600	
	Cash sales	1588	
	Sold goods to Judith Palmer		
	(making special allowance of £225)	1300	gross
23	Paid amount due on van to Roy Fenton	2500	
26	Received payment of Miss Allen's account	600	
30	Paid wages for month in cash	2400	
31	Paid rent for July	360	
	Paid all surplus cash over £300 into bank		

2 Copies of two of David Shaw's personal ledger accounts are shown below.

Bryan Benson

July			£	July			£
14	PRB	13	120	3	PDB	28	960
31	Balance c/d		2040	12	PDB	35	1200
			2160				2160
				Aug.			
				1	Balance b/d		2040

Esther Allen

July				July			
5	SDB	24	600	26	Cheque	CB	600

Which, of the two accounts shown, is that of the debtor? What was the position with regard to both accounts on 10 July and again on 13 July?

19

Final accounts of a sole trader

In this chapter you will learn:
* **_how to prepare the final accounts of a sole trader_**

The meaning of current assets and current liabilities

We have now dealt with all the day-to-day transactions of a business, and we can now bring them together to prepare a set of final accounts for a small trader. If, however, you are still not feeling confident at the basic bookkeeping entries of credit and debit, it is best to go back now and practise some more. From this chapter onwards, **basic bookkeeping is taken for granted, and real accountancy begins!**

Look back at Chapter 18 and see how the entries in the trial balance were split between balance sheet and trading and profit and loss account. The nominal accounts form the trading and profit and loss account; the real accounts go to the balance sheet. The personal accounts are totalled to form debtors and creditors, and these go to the balance sheet as well.

A sole trader's final accounts are now depicted with rather more detail than that we have already seen, followed by explanatory notes.

Trading and Profit and Loss Account
for the year ended 30 June

Turnover		18 500
Less returning inwards		300
		18 200
Cost of sales:		
Opening stock		2 000
Purchases for year	8 000	
less Returns outwards	(200)	
		7 800
add Carriage inwards		120
Warehouse wages		3 380
		13 300
less Closing stock		(2 500)
		10 800
Gross profit		7 400
Commission received		680
		8 080
Salaries & NIC	2 800	
Rates	360	
Advertising	410	
Carriage outwards	130	
General expenses	180	
		3 880
Net profit, taken to capital a/c		4 200

Balance Sheet of A. Trader as at 30 June

Assets employed

Fixed assets

	£	£
Premises at cost	10 000	
Machinery at cost	3 000	
Fixtures/fittings at cost	350	
Motor van at cost	2 650	16 000

Current assets

	£		
Stock 30 June	2 500		
Trade debtors	1 800		
Bank	670		
Cash	30	5 000	

Current liabilities

Trade creditors	1 280		
Expense creditors	20	1 300	
Net current assets			3 700
Net assets			£19 700

Financed by

Proprietor's capital

	£	£
Balance 1 July	19 000	
Net trading profit	4 200	
	23 200	
Less drawings	3 500	19 700
		£19 700

The trading and profit and loss account

Purchases are shown less *returns outwards* to give the net figure of £7800. Carriage inwards (on purchases) increases the cost of the goods bought and in consequence is a trading account expense.

Warehousing costs are considered to be part of the cost of sales, and so are added here. This gives a figure of £13 300. However, we have not sold all of these goods: we still have some closing stock valued at £2500. This has to be deducted to leave the cost of sales figure of £10 800.

On the credit side of the trading account is the net turnover of cash and credit sales, a total of £18 200. The difference between the actual net sales (£18 200) and the cost of those sales (£10 800) is the gross trading profit of £7400.

Most of the profit and loss account debits have come from the cash book or from an expense voucher to be settled later by a cash payment. *Carriage outwards* is an expense of selling and distribution, debited on this lower account to distinguish it from *carriage inwards* (on purchases). *Both are expenses and debits.*

Salaries generally refer to payments made to the monthly paid staff, comprising office and administrative personnel, normally a more static figure than the wages paid on the production and warehousing side. The wages paid to office cleaners, however, should be charged to the profit and loss account.

The few items found on the credit of profit and loss account are usually recognized by the word 'received' and generally refer to small gains or miscellaneous receipts such as 'rents received', 'commissions received' and 'discounts received'.

It is important to remember that neither assets nor drawings appear in the revenue accounts, the only exception being the opening and closing stocks, needed to adjust the net debit for purchases to find the actual cost of the materials consumed during the trading period.

The balance sheet

The *fixed assets* (also known as *non-current assets*) have been listed and totalled first, in order of permanency, usually the case with trading firms.

Premises would be regarded as more permanent and likely to be kept for a longer term than any of the other fixed assets.

Note that the valuation of all fixed assets is at cost, presumably their original cost. This is not likely to be their true worth, as wasting assets such as vehicles and machinery are subject to heavy wear and tear. Their reduction in value (called **depreciation**) is dealt with in Chapter 24.

The *current assets* are generally listed in the order shown, with the least liquid item always first and the most liquid item coming last. These assets change their form in the course of trading; stock is sold to cash and credit customers (the unpaid accounts of the latter being recorded in the books as trade debtors). As the money comes in, further goods are bought for cash or on credit (thus creating trade creditors) and so the trading cycle goes on.

Bank and cash are called 'liquid assets' and petty cash is normally absorbed into the main cash or paid into the bank at the balance sheet date.

Insight

A *current asset* is expected to be able to be turned into cash within 12 months, and is held for trading purposes.

A *current liability* is expected to be settled within 12 months and is being used for trading purposes.

A *fixed asset* (*non-current asset*) is any asset that does not meet the criteria of a current asset.

Long-term liability (*non-current liability*) is any liability that does not meet the criteria of a current liability.

The total current assets here are £5000. From the current assets we have to deduct the current liabilities. These are the amounts we owe which will have to be paid in the near future – typically the next 12 months. In this case they are the trade creditors of £1280 and the expense creditors of £20. A bank overdraft would also be included here, but a long-term loan would not.

The total of current liabilities (£1300) is then deducted from the total of current assets (£5000) to give the *net current assets* of £3700. Net current assets is sometimes referred to as 'working capital'.

If there were any other liabilities these would form a separate entry which follows net current assets, called 'long-term liabilities' – this is where a long-term loan would be entered. However, in our example there are no long-term liabilities. So the current assets are added to the fixed assets to give a total of *net assets*.

The final element of the balance sheet is to show how the capital of the business is owned. In a sole trader's accounts this is straightforward; it is owned by the sole trader. The opening balance brought forward from last year is £19 000. The profit from the profit & loss account of £4200 is added to this; if there was a loss it would have to be deducted. However, the trader has also taken £3500 out of the business as drawings, so this needs to be deducted.

The capital account of A. Trader shown on his balance sheet is simply a summary of his personal account kept in the private ledger, thus:

Dr.   **Capital Account**   Cr.

June			£	July			£
30	Drawings for year		3 500	1	Balance b/f from last year		19 000
30	Balance c/d		19 700				
				June	Net trading		
				30	profit from P&L a/c		4 200
			23 200				23 200
				July			
				1	Balance b/d		19 700

It will be seen that A. Trader has increased his capital (and the net assets) by £700 during the past trading year. This amount is the difference between the profit made and the sum total of his drawings.

PAUSE FOR THOUGHT

Make up a small trading and profit and loss account, inserting your own figures, to include:

> sales, purchases, returns inwards and outwards, stocks at the beginning and at the end of the period, warehouse wages, carriage inwards and outwards, rents received, transport expenses, rates paid, salaries, postages, insurances, heating and lighting.

Show the cost of sales in the trading account.

Following the drafting of the revenue account on the above lines, continue with a balance sheet showing the layout similar to the illustration at the beginning of this chapter. Show the working capital and the net assets in completion of the assignment.

Insight

Net assets means the difference between the total assets and the total liabilities of the business; it represents the amount of the owner's equity in the business. The owner's equity may be increased by earning profits or introducing more capital. The owner's equity can also decrease by making losses or withdrawing some of the capital.

KEY POINTS

▶ The one main and useful equation in accounting is simple enough to remember:

$C + L = A$

where

C is the proprietory capital; (also known as owner's equity)

L is the external liabilities of the business (current liabilities + non-current liabilities); and

A is the sum total of the assets and property of the business (current assets + fixed assets).

When working out problems of capital, net worth and equity, given any two of these basic business elements, the third can easily be calculated.

▶ In all these elementary exercises, the proprietor's capital account on the balance sheet should be presented clearly. In no circumstances should either trade or expenses creditors be included in the capital section.

TESTING YOURSELF

19.1 Refer back to Revision Exercise 2 and make up the trading and profit and loss account for the month and a balance sheet as at 31 July from the trial balance already extracted, taking the valuation of Mr Shaw's closing stock at 31 July to be £450.

19.2 What financial statements do the nominal accounts, personal accounts and the real accounts go to?

19.3 Make up the trial balance and a complete set of accounts, as at 30 June, from the following balances extracted from the financial books of Lucian Lane, a small retail merchant whose business is mainly cash with a few credit sales.

Show clearly on his balance sheet:

a *separate headings for fixed and current assets, the capital, and current liabilities.*

b *Mr Lane's working capital and his net assets at 30 June.*

	£
Capital account 31 May	2000
Carriage inwards	42
Cash balance 30 June	67
Bank balance 30 June	1850
Carriage outwards	56
Warehouse expenses	235
Petrol and oil	86
Fittings and fixtures	340
General expenses	28
Returns inwards	66
Returns outwards	44
Van repairs	52
Stock 31 May	320

Purchases	3682
Sales	8760
Insurance	34
Advertising	44
Rent and rates	420
Stationery	32
Trade creditors	220
Drawings	800
Salaries	1600
Trade debtors	430
Van at book value	840

Mr Lane valued his closing stock at £480.

20

Interpretation of accounts

In this chapter you will learn:
- *how to calculate profit and stock ratios*
- *how to calculate liquidity ratios*
- *how to calculate gearing ratios*
- *what these ratios can (and cannot) tell you about a business*

Users of financial statements

There is a variety of different users of financial statements. This includes, for example, the owner, existing and potential shareholders, creditors including the bank and suppliers, employees, and the general public. They all will have different reasons for wanting to review a business's financial statements, for instance the bank will need to know how risky the business is in order to ascertain whether or not to grant the loan, and the employees may want to know whether or not the company is prosperous. The calculation of ratios can aid the interpretation of the financial statements for these users.

The trading account determines the gross profit (or loss) over a defined period of trading, providing useful information to management, in particular when comparisons are made with the trading accounts of other periods, or with those of competitors.

To arrive at some of the bases of comparison, namely the cost of sales, gross profit percentages and the rate of stock turnover, look at the following simple and old-fashioned horizontal style of trading account:

Trading Account for the month of January

	£	£		£	£	£
Stock 1 Jan.		800	Cash sales		450	
Purchases	5000		Credit sales	5800		
Less returns	300	4700	*Less* returns	250	5550	6000
		5500				
Less stock 31 Jan.		1000				
Cost of sales		4500				
Gross profit		1500				
		6000				6000

COST OF SALES

The cost of sales (or the cost of the goods sold), £4500, is found by adding the opening stock to net purchases and deducting the figure for closing stock. This cost of sales figure is not the same as the goods bought or purchases, but the valuation of closing stock probably includes much of the merchandise recently bought.

The cost of sales figure is also increased by other trading account debits such as wages, carriage inwards and warehousing expenses, omitted here for simplicity.

Gross profit and mark-up percentages

The trading results of one period may be compared with those of another period (or of a competitor) by expressing their gross profits as percentages of their relative sales turnovers. In the trading account illustrated the percentage is arrived at thus:

$$\frac{\text{Gross profit}}{\text{Net sales}} \times 100, \text{ that is } \frac{1500}{6000} \times 100 = 25\%$$

Whether this is a fair gross profit percentage will depend upon the nature of the business.

If there were a wide variation between the gross profit margins of one period and the next (and it is not a seasonal business) management would investigate the reason for the discrepancy.

This percentage is different from the one which is used by the trader to arrive at a selling price, known as the **mark-up**. The mark-up percentage is the amount by which the cost of the goods *to the trader* is to be increased to arrive at the selling price, i.e. gross profit over cost of sales. In the above example it would be:

$$\frac{1500}{4500} \times 100 = 33.3\%$$

Rate of stock turnover

The number of times the stock is 'turned over' in the course of a month or a year has a direct bearing upon the gross profit of a business. Generally, profit increases if the rate of stock turnover can be improved and the cost of sales remains fairly proportionate to the turnover.

The rate of stock turnover is calculated by dividing the *cost of sales by the average stock*.

The average stock in the illustration is: $\dfrac{800 + 1000}{2} = £900$

The rate of stock turnover is then:

$$\frac{\text{Cost of sales}}{\text{Average stock}} \text{ that is } \frac{4500}{900} = 5 \text{ times in the month.}$$

In general, the faster stock is being turned over the better, because it reduces the amount of working capital needed to keep the business going. However, it is not realistic to compare stock turn between businesses operating in different markets. A food retailer will turn over the stock within a matter of a few weeks at most; an antique dealer may need to keep a very large amount of stock on hand in order to give customers an adequate choice.

More relevant is a comparison of stock turn for different periods within the same business. If the stock turn is going down this means that stock is growing. Unless this is accompanied by a significant increase in sales, it probably means that stock is simply piling up unsold. It will still have to be paid for, however, and this may lead to the business running out of money.

Comparison of business profits

From the information tabled below you are asked to give your opinion, supported by percentages, as to which business (operating in the same trade) seems to be the more progressive of the two.

	Capital employed £	Turnover £	Gross profit £	Net profit £
Business A	7500	6000	3000	1500
Business B	10 000	8000	3600	1800

At first glance, the second business may appear to be more progressive, because it is making more profit. However, the greater gross and net profit of Business B are based upon its greater (by one-third) capital and turnover than Business A, leading then to the natural assumption that the gross and net profits of Business B should be even larger than those shown.

Calculating profit percentages highlights the problem.

a *Percentage of gross profit to turnover:*

Business A $\dfrac{3000}{6000} \times 100 = 50\%$

Business B $\dfrac{3600}{8000} \times 100 = 45\%$

b *Percentage of net profit to turnover:*

Business A $\dfrac{1500}{6000} \times 100 = 25\%$

Business B $\dfrac{1800}{8000} \times 100 = 22.5\%$

c *Percentage of net profit to capital employed:*

Business A $\dfrac{1\,500}{7\,500} \times 100 = 20\%$

Business B $\dfrac{1\,800}{10\,000} \times 100 = 18\%$

This comparison by percentages reveals that Business A, though operating on a smaller capital, is the more progressive business, provided, of course, that the figures given for both businesses are truly representative for their trade, with abnormalities excluded.

Liquidity ratios

One of the reasons for looking at the accounts of a business may be credit control – seeing whether it is appropriate to extend credit to it for goods purchased. In this case it is not so much the profitability of the business that matters but its ability to pay its debts as they fall due. This in turn means that it must have readily realizable liquid assets. Two main ratios are used for considering whether this is so.

▶ *The* **current ratio** *is the more traditional of the two, comparing all current assets (generally stock, debtors and cash) over current liabilities. It is often said that this ratio should be over 2 – i.e. current assets should exceed current liabilities by the ratio 2:1 at least. However, the problem with this ratio is that it treats a business as liquid even if it has a high level of stock which simply cannot be sold.*
▶ *The alternative ratio, and the one often relied on more for credit decisions, is the* **acid test ratio.** *This is the same as the current ratio, but excluding stock. It is often said that this ratio should be at least 1, on the basis that it can then pay its creditors as they fall due out of the money it receives from its debtors.*

Care should be taken in interpreting these ratios too strictly. An expanding company may well have ratios which look adverse but, provided it has sources of working capital (such as a bank overdraft facility still unused), it will still be liquid and worth doing business with. The ratios need to be looked at in the context of what is normal for the business, and attention paid to changes from one year to the next.

Gearing ratios

Gearing is a measure of the financial risk of the business. It describes the mix of loan finance and equity finance in a company; how much of the money invested in the business has come from borrowed funds and how much from non-borrowed funds.

It is calculated by:

non-current liabilities/owner's equity + non-current liabilities.

A high gearing percentage indicates a high exposure to financial risk, where the business will have interest charges to be met as well as repaying the loan when it is due. Therefore another useful ratio here to calculate is interest cover:

operating profit (before interest and tax)/interest

This ratio shows if the profit generated by the business can cover the interest charges sufficiently. There will be increased cause for concern if the interest cover is low.

Insight

Ratios have limitations which are important to bear in mind.

▶ *Ratios taken in isolation for a single company or period of time have very limited usefulness.*
▶ *No two businesses are exactly alike therefore comparisons must make allowances for this.*
▶ *Different businesses have varied policies on valuations of different items such as stock. This would cause differences.*
▶ *Ratios are only a guideline to help highlight areas that require further investigation.*

KEY POINTS

▶ In problems concerning cost sales or stock turn, always make up a trading account showing cost of sales.
▶ A critical appraisal of trading account figures might invoke these questions:
  ▷ Are the figures as good as expected?
  ▷ Are the figures as good or better than those of the previous year?
  ▷ Are certain lines more profitable than others?
  ▷ Should economies be made in some departments?
  ▷ Is money or stock being embezzled and what precautions should be taken?

TESTING YOURSELF

20.1 The balances below were extracted from the books of Norman Deal on 31 January. Find his gross and net profit percentages on turnover; also his rate of stock turnover.

	£		£
Stock 1 Jan.	600	Net purchases	2300
Stock 31 Jan.	900	Net sales	6250
Warehouse expenses	205	Salaries	430
Carriage inwards	45	Advertising	80
Carriage outwards	50	Wages	1500
Rent and rates	400	General exes.	40
Heating/lighting	160	Comm. rec'd	535

20.2 Say which, in your opinion, is the more profitable of the two businesses, from the figures given below. Give your reasons.

	Average stock	Stock turn	Mark-up on cost
	£		
Business RS	2000	6.5 times	40%
Business KC	2000	8.5 times	32%

20.3 The balances below were extracted from the books of Razi Ahmed on 31 March. Calculate his interest cover.

	£
Gross Profit	5000
Rent and rates	350
Heating/lighting	200
Interest charges	2000

Revision exercise 3

Gregory Bartel's balance sheet at 30 June was made up on these lines:

Assets employed			
Fixed assets at cost	£	£	£
Premises	5000		
Fittings	300		
Motor van	1800		7100
Current assets			
Stock	900		
Trade debtors	696		
Bank	1250		
Cash	54	2900	
Less: Current liabilities			
Trade creditors	825		
Rates owing	175	1000	1900
Net assets			9000
Long-term liabilities			
Loan from Mrs G. Bartel			3000
Net assets			6000
Financed by:			
proprietor's capital			6000

Your assignment is to open a new set of account books from the information given above, and then post up Gregory Bartel's trading transactions for the month of July, as listed below from the original entry stage.

The trade creditors comprised two suppliers' accounts, viz. Hugo Jones £550 and Mohammed Iqbal £275, totalling £825 on the above balance sheet.

The trade debtors' total of £696 comprised four accounts, namely:

Hilary Fane	£180	*Delia Harlow*	£162
Petula Hill	£144	*Richard Mead*	£210

Balance up all accounts and take out a trial balance on 31 July. Make up the trading and profit and loss account for Gregory Bartel for the month of July, and a balance sheet as at 31 July, accepting his own valuation of closing stock at £860.

Transactions for July:

July

1   Bought £160 of goods from Hugo Jones, being allowed 25% trade discount
     Cash sales £84

2   Paid creditors for rates £175
     Cash sales £141; paid £100 into bank

6   Drew cash from bank and paid wages £150
     Miss Fane sent cheque for £180 in settlement of her June account
     Paid £300 'on account' to Hugo Jones
     Bought £80 desk for office; paid by cheque

9   Cash sales £165; paid £100 into bank
     Paid advertising account £20, and garage account £25

12   Daily takings amounted to £137
     Paid £150 wages out of cash
     Repaid Mrs Bartel £500, reducing loan account to £2500

14   Received cheque for £162 from Delia Harlow. She bought another £60 of goods. Allowed her 20% trade rate.

15   Cash sales £159; cash purchases £18

18   Paid wages £160
     Sold £80 of merchandise to Petula Hill less 10% trade rate:
     She paid £100 off her outstanding account.

(Contd)

23	Bought £240 of goods from Hugo Jones, being allowed the usual discount. Also paid £15 delivery charges in cash.
	Bought £120 of goods from Mohammed Iqbal who would only allow £10 trade discount. Gave him a cheque for £275 in settlement of his June account.
25	Returned £40 (gross) merchandise bought from Hugo Jones on 23 July
	Cash sales £136; paid £160 wages out of cash
28	Sold £75 of goods to Dinah Mortimer allowing her 20% trade discount. Paid £6 carriage on various deliveries made during the month of July.
30	Cash sales £128
	Withdrew £100 cash for private and family use
31	Paid £80 cheque for design services.
	Retained £25 cash for office and paid surplus cash into bank.

21

Cash flow statements

In this chapter you will learn:

- *the difference between profit and cash flow*
- *how to reconcile them*
- *how to create a cash flow statement*

Introduction

In a very simple trading situation, the movement of cash is exactly the same as the profit or loss. Say a market trader buys flowers for her Saturday market stall costing £300 and sells them for £600 (throwing away any she cannot sell, as they will not last until the next week's market). She also pays £40 on the day for the hire of the market stall. Her profit is obviously £260, and the amount of cash in her pocket has also increased by £260.

However, it only takes a slightly more complex business for the cash and profit and loss account position to diverge. If the florist in the example above was in a market that opened every day, then stock could probably be kept for a day or two. If she had £50 of stock at the end of the day, her cash would still have increased by only £260 but her profit would have increased by £310, as the cost of sales would be £300 of purchases less £50 of stock.

While the balance sheet and profit and loss account will normally give a better picture of how a business is doing, it is clearly true that control of cash is an important issue in a business, and seeing what has happened to cash during the year can sometimes throw light on the accounts. Larger organizations are therefore generally required to produce cash flow statements as part of their financial statements.

> ### Insight
> Cash flow statements are not always in the syllabus for first-level examinations such as GCSE. If you are using this book to prepare for an exam, now is a good time to check the syllabus.

Cash flow

PURPOSE OF A CASH FLOW STATEMENT

The *starting point* for a cash flow, or funds flow, statement is the reported net profit or loss. The *finishing point* is the total increase or decrease in cash and bank balances. Both of these figures are of course easily identifiable in the balance sheet. The purpose of the cash flow statement is to reconcile the one to the other. This is normally done in three sections:

▶ *cash flows from operating activities*
▶ *cash flows from investing activities*
▶ *cash flows from financing activities.*

These are considered in turn shortly.

In making these adjustments, it is important to remember which way the adjustment needs to go. The statement starts with profits and ends with cash. Anything that increases profits but does not increase cash must therefore be *subtracted*, anything which reduces profits but does not reduce cash must be *added*. Similarly, anything that increases cash but does not increase profits must be *added*, anything which reduces cash but does not reduce profits must be *subtracted*.

CASH FLOWS FROM OPERATING ACTIVITIES

As the title suggests, this is the section where trading adjustments are detailed. The main examples of this are:

▶ **Changes in stock** *Increases in stock will increase profits (by reducing cost of sales) but will not increase cash, so if stock has gone up the increase must be deducted in the computation, if it has reduced it must be added.*

▶ **Changes in debtors** *An increase in debtors means profits increased but no increase in cash, so if debtors have gone up the difference must be subtracted, if debtors have gone down it must be added back.*

▶ **Changes in creditors** *This is the reverse of debtors, so if creditors have gone up the difference must be added back, if they have gone down the difference must be subtracted.*

These adjustments will be required in virtually all cases. The following three adjustments may not be quite so all-pervasive, but will be found in a majority of statements.

▶ **Depreciation** *Because this is a book entry that reduces profits but has no cash impacts it has to be added back in the statement.*

▶ **Loss/profit on sale of fixed assets** *Because this is essentially an adjustment for insufficient or excess depreciation when you dispose of a fixed asset, a loss is added back and a profit is deducted. The actual proceeds from the sales are dealt with under investing activities.*

▶ **Other provisions** *Where provisions have been made for possible expenditure not yet incurred, they must be cancelled out in the cash flow statement. The most common example is a bad debts provision, against debts that are still in the debtors' total but where payment is unlikely. So if, as is normal, a provision has resulted in a net expense in the profit and loss account, it must be added back in the cash-flow statement. Rather less likely, a provision made in a previous year may have been reduced, resulting in a net addition to profits. This must be subtracted.*

CASH FLOWS FROM INVESTING ACTIVITIES

Essentially these are transactions affecting the fixed assets part of the balance sheet. Whilst the profit or loss on a sale of fixed assets has now been adjusted for, the fact that *what was a fixed asset is now cash* has not. This increases cash without increasing profit and must therefore be added. The main adjustments are:

- ▶ **Sale of equipment etc.** *As detailed in the previous paragraph, the cash received for such a sale must be added in the statement.*
- ▶ **Purchase of equipment etc.** *Correspondingly, a purchase of equipment will have reduced the cash balance of the organization without having had a direct effect on profit. This must be deducted in the statement.*
- ▶ **Investments** *The principles for investments are similar; a purchase of shares in another company (for example) is a reduction in cash and must be subtracted in the statement, whereas a sale of an investment increases cash without increasing profits, and must be added. Adjustments for investments must be distinguished from financing, covered below. Investments are made by the business in other organizations, whereas financing is the investment made by others in the reporting organization.*

CASH FLOWS FROM FINANCING

The most immediate form of bank financing, the overdraft, will show up automatically in the cash flow statement when it finishes with cash and bank balances. However, all other financing movements need to be accounted for, including the return on financing received by those who provide it.

- ▶ **Drawings** *This is the return that a proprietor of an unincorporated business receives. Since it is a reduction of the cash in the business that does not affect profit, it has to be deducted.*
- ▶ **Capital introduced** *If capital is brought into the business, whether from the proprietor or a third party, the cash balance*

will increase without any corresponding increase in profit, so this has to be added back.

▶ **Capital repaid** *Most often this will be the repayment of a loan, but in partnerships it can be the repayment of partnership capital. As the opposite of capital introduced, it has to be subtracted in the calculation.*

▶ **Dividends and interest on loans** *Similar in concept to drawings, dividends and interest on loans represent the return that investors expect on their capital invested in the business. As such they reduce cash without affecting profit, and must therefore be subtracted in the calculation.*

Preparation of a cash flow statement

Now that the adjustments to be made are understood, we can look at the preparation of a cash flow statement. Given a balance sheet and profit and loss account, with comparative figures, the statement should be straightforward to prepare, and has an inbuilt check since the calculation should end with the change in bank and cash balances as shown.

Example

Look ahead to Chapter 26, and you will find a balance sheet and profit and loss account for Robert Morris. Before preparing the cash flow statement, you need some more information about the previous year's balance sheet, which would be in the column headed 'last year's figures'. The figures you need are these (some of them are referred to in the profit and loss account):

Stock	*£1200*
Debtors:	*£2000*
less a provision of £120	*(£120)*
Cash at bank (not overdrawn):	*£2065*
Rates in advance:	*£350*
Cash in hand:	*£55*
Trade creditors:	*£1000*
Expense creditors:	*£750*

The only change to fixed assets from the previous year to the current year is that machinery and equipment were purchased during the current year for £3000. The advertising expenditure carried forward was also £200 last year.

The cash flow statement is as follows.

Cash flow statement for Robert Morris			
Cash flow from operating activities:	£	£	£
Net profit		2890	
Increase in stock	(250)		
Depreciation	760		
Decrease in debtors	200		
Reduction in bad debt provision	(30)		
Increase in prepayment	(50)		
Increase in trade creditors	400		
		1030	
Cash flow from investing activities:			
Purchase of equipment		(3000)	
Cash flow from financing:			
Drawings		(3745)	
Decrease in cash			(2825)
Bank balance at end	(730)		
Cash balance at end	25		
		(705)	
Bank balance at start	2065		
Cash balance at start	55		
		(2120)	
Decrease in bank/cash balances			(2825)

22

Capital and revenue expenditure

In this chapter you will learn:
- *the difference between capital and revenue expenditure*
- *the effect that each has on the final accounts*

The term **capital** so far has been used to indicate the personal capital accounts of sole traders.

Sometimes, though, the term is used to describe capital assets acquired for the permanent (or at least multiple year) use of the business, and the expenditure used to buy those assets is then known as capital expenditure. This would, for example, include land and buildings, plant and machinery, furniture and fixtures, motor vehicles and office equipment.

On the other hand, **revenue expenditure** refers to the cost of maintaining and operating these capital assets, including all repairs. Revenue expense also embraces all costs of maintaining sales revenue and all normal trading expenses, such as the payment of wages and salaries, materials bought for re-sale or conversion, advertising, insurance, heating and lighting, repairs and transport of all description, in fact all those debits to be found on the trading and profit and loss account.

Capital expenditure

A new business starting from scratch might have only a sum of money in the bank as its sole asset. In the course of trading, various fixed assets would be acquired, probably some fixtures and fittings, some machinery and perhaps a motor van. The payment for these capital assets would be made either from the original money capital, or from profit surpluses retained by the business as it trades from year to year. This is capital expenditure.

Again, these (or similar) assets could be included in the purchase price when a business is bought as a 'going concern', and any further payments made on improving or extending these assets would be debited to these same accounts and increase their 'book value'. See the illustration below which shows how fixed asset accounts may appear in a small business's accounts.

Dr.   **Machinery Account**   Cr.

			£				
Jan.1	Balance b/f		4500				
May 5	Cheque (new machine)	CB	1200				

Furniture and Fixtures

			£				
Jan.1	Balance b/f		720				
Oct. 15	Cheque (filing cabinet)	CB	40				

Vans Account

			£				
Jan.1	Balance b/f (second-hand van taken over)		1800				

At the end of the financial year, the balances of these asset accounts would be shown on the balance sheet under the heading of *fixed assets*, thus:

Assets Employed		
Fixed assets	£	£
Machinery at cost	5700	
Fittings at cost	760	
Vans at cost	1800	8260

Revenue expenditure

Revenue expense ends up as a debit to the trading account or the profit and loss account at the end of the financial year. The illustration below shows how these might appear in your books and summary report.

Dr.   **Warehouse Expenses**   Cr.

			£			£
Mar. 4	Cheque for re-decorating, etc.	CB	420.00	Dec. 31	Transfer to trading account	498.00
May 12	Cleaning	CB	24.00			
Aug. 6	Cleaning	CB	24.00			
Nov. 22	Cleaning	CB	30.00			

Advertising

			£			£
				Dec. 31	Transfer to P&L a/c	68.70
Jan. 6	Cheque	CB	15.50			
Apl. 3	Cheque	CB	28.20			
Oct. 15	Cheque	CB	25.00			

Van Repairs and Renewals

			£				£
Mar. 25	Red Garage (service & new exhaust)	CB	68.50	Dec. 31	Transfer to P&L a/c		120.90
Sept. 2	Service, etc.	CB	52.40				

Trading and Profit and Loss Account
for period ended 31 December

	£		
Warehouse expenses	498.00		
Gross profit b/d	XX		
Advertising	68.70	Gross profit c/d	XX
Van repairs/renewals	120.90		

Insight

Just in case you missed this: note capital expenditure is in the Balance Sheet while revenue expenditure is shown in the Profit and Loss account.

Money received from the sale of capital assets is debited to the cash book and *credited to the related asset account*. Receipts of this nature are not regarded as revenue income like sales and cash takings. Remember that when an old asset has been sold and its book value must now be erased from the accounting records. This means that any balance remaining on the account (normally a debit balance) must be written off to the profit and loss account, as it is a loss and expense suffered by the business. There is often then also an adjustment in connection with its depreciated book value if the asset has been subject to a depreciation charge. We will discuss how this works in practice in Chapter 24.

The distinction between capital and revenue expenditure is important, as the charging of heavy capital outlay against profits would incorrectly reduce (often substantially) the net profit and the tax liability of the business.

Revenue Expense	Capital Expense
is debited and charged to the profit and loss account, decreasing the net trading profit.	is debited to the asset account and taken to the balance sheet under the heading of fixed assets. Capital expenditure must not be charged against profits.

KEY POINTS

▶ The term 'revenue' refers to either income or expense having a direct connection or influence on the trading returns or profits of the business.
Revenue expense is charged and debited against profits, whereas revenue income implies additional profit or gain.

▶ Money spent on repair work and renewals (including re-decoration) is revenue expense and is debited to the profit and loss account.
Money spent on improvements or extensions to existing capital assets is capital expenditure and must be debited to the asset account.

▶ In examination problems on capital and revenue expenditure, ask yourself: Has the money been used up in the ordinary course of trading (rent, stationery, wages, etc.), or will its use and benefit extend perhaps over a number of years (as in the case of a new van, office equipment or a building extension)?
Another point about a building extension or the erection of machinery is that the wages paid to the firm's own workers engaged on the erection work is also capital expenditure to be debited to the cost of the asset.

TESTING YOURSELF

22.1 Re-draft this trading and profit and loss account of Miranda Moss, incorrectly drawn up by her younger brother.

Trading and Profit and Loss Account as at 31 August

	£		£
Stock 31 July	680	Sales	7200
Purchases	3440	Rets. Inwards	60
Rets. outwards	26		
Carriage outwards	18	Stock 1 July	550
Wages	520		
Salaries	1240		
Gross profit c/d	1886		
	7810		7810
Stationery	15	Gross profit b/d	1886
Drawings	360	Loss on sale of van	130
Van repairs	45	Sundry receipts	68
Rents received	84	Carriage inwards	42
Second-hand van	830	Net loss on trading	54
Sundry expenses	26		
Advertising	24		
Answering machine	96		
Showroom extension	580		
Decorating office	120		
	2180		2180

23

The general journal

In this chapter you will learn:
- *how to use the journal to calculate opening capital*
- *how to use the journal to correct errors*

We have already used journals (day books) in connection with the original entries for credit purchases and credit sales. Another subsidiary book, the **general journal**, is now brought into use as a list of transactions of a special or unusual nature which cannot conveniently be taken to another book of original entry.

> **Insight**
>
> In the UK, the general journal does not form part of the double entry system, but is used as an aid or a guide to help the ledger accounting staff with their postings. Sometimes examination questions will ask for an answer in journal form, even though it would not actually be dealt with in this way by a bookkeeper. The advantage of asking for the answer in this form is that it provides a structured and succinct way to check whether the student understands the question.

Opening entries

In some examination problems the proprietor's capital at the start of the exercise is not given. The student may be expected to find the opening capital of the owner/proprietor from a list of assets and liabilities such as the following:

> **Fittings £250; Machinery £1700; Opening stock £1400;**
> **Trade debtors: Jones £245 and Brown £105**
> **Trade creditors: Evans £65 and Smith £210**

In addition to the assets and liabilities listed above, Frank Lawson, new owner of this business, after settlement of all commitments with vendor, has the sum of £575 in the business bank account.

The first stage in this type of problem is to list the opening assets and liabilities in the journal in trial balance style, thus:

Opening Entries 1 July

	Debit £	Credit £	
Fittings	250		
Machinery	1700		
Stock on hand	1400		
Bank	575		
Debtors: Jones	245		
Brown	105		
Creditors: Evans		65	
Smith		210	
Capital F. Lawson		4000	← the difference or balancing figure
	4275	4275	

The opening capital of Frank Lawson is worked out by addition or subtraction to give the balancing figure of £4000, also confirmed by the simple accounting equation of $A = C + L$, where A represents the assets, C the proprietor's capital, and L the external liabilities of the business.

If this problem formed part of an exercise, the asset balances would be posted to the debit of their respective accounts in the cash book, the sales ledger and the general ledger; the balances owing to the two creditors would be posted to the credit of their accounts in the bought ledger. In this manner Frank Lawson's new set of accounting books would be put on a proper double entry basis as from the beginning of the trading period commencing 1 July.

It is not usual to total journal entries except in the case of opening entries as shown in the illustration (p. 178). From the examples below, it will be seen that these details are merely recorded in the journal so that the ledger clerks will understand the reason for the entries and be able to post the debits and credits to their correct locations.

Insight

Note the orthodox method of journalizing by recording the debit entry first in the left-hand column, with the credit entry underneath in the right-hand column, the wording slightly indented to the right of its upper counterpart.

The brief explanation of the journal entry, typed or written before ruling off the entry, is called the **narration**.

Trial balance errors

The first check on the accuracy of the books is that the trial balance totals should agree. However, there are errors which do not throw the double entry out of balance, but which still need to be corrected when they are identified. Here are five of them.

► **Error of omission** *This occurs when goods are bought or sold on credit, but the original documents in connection with the purchase or the sale are forgotten or mislaid. In consequence, neither the debit nor the credit entries are posted in the books.*

► **Error of duplication** *Here, the original records become duplicated and the customer is charged double the amount that is actually owing. Normally, this is found and rectified fairly quickly once the monthly statement of the supplying firm has been sent out.*

► **Error of commission** *This occurs when the amount, usually a payment, is debited or credited to the account of the wrong person. For example, a payment of £8 was received from T.H. Johnstone but was incorrectly credited to the account of T.N. Johnson, and has now been adjusted through the journal.*

► **Compensating error** *Mistakes in arithmetic of similar amounts, often round figures of £10 or £100 commonly occur in additions, subtractions or carry-forwards, in particular in the day book and when the accounting system is not computerized. For example, the bought day book may have been totalled to £150 more than it should be in a carry-forward, and an error also made of £100 on one of the big debtors' accounts with the balance at £100 less than it should be. This would be corrected through the journal by a debit of £100 to the debtor's account and a credit of £100 to purchases account.*

► **Error of principle** *An amount may have been posted to an expense account instead of to an asset account, or vice versa. Again, see the example below, where the machinery repairs account is being debited, through the journal, with £15 which in the first instance has been wrongly charged to the asset account. This is an important aspect of capital and revenue expenditure.*

The General Journal

			£	£
	Special entries			
March				
5	Cash	Dr.	25	
	Fittings/Fixtures a/c			25
	Being sale of old cupboard			
June				
30	Bad debts account	Dr.	14	
	Alan Jones			14
	Irrecoverable balance written off to Bad Debts a/c			
	Correction of errors			
July				
9	T.N. Johnson	Dr.	8	
	T.H. Johnstone			8
	Payment from T.H. Johnstone had been incorrectly posted to the credit of T.N. Johnson			
Sept.				
2	Machinery Repairs a/c	Dr.	15	
	Machinery (asset) a/c			15
	Adjustment of item wrongly charged to asset account			
	Year-end transfers			
Dec.				
31	Trading Account	Dr.	9100	
	Stock at 1 Jan.			270
	Purchases for year			8640
	Carriage inwards			190
	Transfer of year-end balances			

PAUSE FOR THOUGHT

Describe the function of the general journal and the purpose in this country for which it is used, more as an aid to the bookkeeper/cashier than as an integral part of the double entry system. In what way does the general journal differ from the bought and sales journals?

Of the five kinds of trial balance errors which might, for a while, remain undetected, say in which category you would place the following:

▶ *A £100 error in the carry-forward of the bought journal, offset by £100 error in the addition of the cash sales total*
▶ *Wages of £400 paid to own workers in erecting their new welding machine, the expense payment being debited to trading account*
▶ *George Green has been invoiced £15 for goods ordered and sent to Miss Georgina Green.*

KEY POINTS

▶ The necessary requisites of a journal entry are:
  ▷ The date is entered in the journal.
  ▷ The account to be debited is first shown, with the amount in the first column.
  ▷ The account to be credited is shown underneath, indented slightly to the right, with the amount in the second (credit) column.
  ▷ A brief explanation of the purpose of the entry, called the narration, is given before ruling off. With the exception of opening entries, totals are not necessary.
▶ In examination problems, where errors are to be corrected through the journal, you are told that *something is wrong,* that a mistake has been made, and you are asked to put it right.
▶ All that is necessary is to *reverse the whole or part* of the incorrect posting already made. As long as you can understand what is wrong, it will be simple enough to put it right.

(Contd)

▶ Some answers only need a single-sided correction. For example, the sales day book has been over-added by £100 to show an excess of £100 on the sales account. Instead of altering the ledger account, a single-sided journal entry is made on the debit of the journal.

TESTING YOURSELF

23.1 Write up the opening journal entries of Wensley Dale's business from the information given below, and ascertain his commencing capital at 1 January.

	£		£
Premises	6000	Mortgage loan of	5000
Fittings	550	(secured on premises)	
Stock	1010	Machinery at cost	2500
Bank	865	Van valued at	1750
Cash	35	Rates owing	290

Trade debtors:
Leo Fraser £425
Abe Fisher £160

Trade creditors:
£320 Lesley Mansel
£185 Petra Nutkins

23.2 Journalize the following.

March

2	Bought second-hand computer and paid by cheque £175
5	Cash purchases £138
12	£52 of goods sold to Anthony Fircone
15	Bought £130 of goods from Horace Heap
20	Sold old computer for £45 cash
22	Paid £5 for stationery
27	Drew £50 cheque for 'self'
31	Transferred opening stock of £950 to trading at account month end

23.3 Make the necessary corrections, through the general journal, of the following.

July

1 A new machine bought for £250 has been debited to purchases account.
3 Sales of £38 on credit to Sue Palliser have been wrongly charged to her sister Sarah.

(Contd)

6 A cheque for £20 from A. J. Smith has been credited to the account of his cousin J. A. Smith.

12 Drawings of £25 were found debited to salaries account.

15 Repairs to machinery £75 have been debited to machinery (asset) account.

18 The purchase of a filing cabinet (£80) has been charged to stationery account.

28 The total of the purchases day book for June was found to have been posted to purchases account as £986 instead of the true figure of £968.

30 Bank interest received at end June £8.50 was debited to bank charges account.

Revision exercise 4

Perdita Gale has a small trading business. Her assets and liabilities are listed at 1 March:

	£		£
Warehouse premises at cost	80 000	Stock at 1 March	9 200
Fittings/fixtures at cost	3 500	Bank balance	4 750
Loan from bank (secured on premises)	40 000	Cash in hand	550

At this date her trade debtors comprised three accounts:

Maud Foster £480   Judi Birch £600   Melanie West £420

Money was owing on three suppliers' accounts, namely:

Giles Stevens £800   Keith Owen £300   Isaac Jacobs £1400

You are required to open a new set of books and post up Miss Gale's assets and liabilities, including her opening capital. Then you are to post up all the listed transactions for the month of March, make up her trading and profit and loss account for the month, and draw up a balance sheet as at 31 March.

Miss Gale valued her closing stock at 31 March at £8500.

March
2   Cash sales £780
     Sold goods £350 to Judi Birch less 20% trade discount.
     She paid her February account less 5% cash discount.
4   Cash sales £860; paid £500 into bank.
     Miss Birch returned £50 gross purchases of March 2.
6   Paid part-time assistant wages £500 in cash and applied to
     HM Revenue and Customs for her PAYE tax code number.
     Paid stationery £40 and stamps £20 in cash.

(Contd)

9	Bought £600 of goods from Isaac Jacobs less 25% T.D. Paid £50 carriage on purchases. Settled Jacobs' February account less 5% cash.
11	Returned £120 (gross) merchandise to Isaac Jacobs. Bought £850 filing cabinets; paid by cheque.
14	Sold £720 of goods to Maud Foster. She paid her old account. Also sold £380 of goods to Melody Robson and opened up new credit account.
15	Cash sales £1260. Cash purchases £60. Paid delivery charges on routine credit sales £30 cash. Paid wages £500.
19	Bought £800 of merchandise from Giles Stevens and settled his February account less 5% for cash. (Note: 'for cash' often implies 'by cheque'.)
22	Received credit note for £120 from Giles Stevens for trade discount on the purchases of 19 March.
25	Cash sales £1400. Paid £500 wages. Miss Gale withdrew £1000 cash for private use.
30	Paid local secretarial agency £600 for temp.
31	All surplus cash above £200 paid into bank.

24

Depreciation of fixed assets

In this chapter you will learn:
- *what is depreciation*
- *why is it important for correct accounting*
- *two methods of depreciation*
- *how to use a provision for depreciation account*

Machinery and vehicles lose value through wear and tear and the passage of time – it is rare to sell assets after use for more than you acquired them for. In business this loss in value is regarded as an expense in the same way as any other routine expense of the business. Money has been paid out which will never be recovered, so this must be reflected in the trading profit of the business for this loss in value of a fixed asset. This loss is called **depreciation**. It is a common (but mistaken) assumption that the depreciated value of an asset should approximate to its second-hand value. The purpose of depreciation is to allocate the total capital outlay over the useful life of the asset, so that the reported profits accurately reflect the true costs of earning them.

Most fixed assets (the big exceptions being land and buildings) are subject to permanent loss in value due to wear and tear (vehicles), the passage of time (leases), obsolescence (changes in method, style or fashion), or simply through general wastage and being used up or worked out in the case of mines. As such, many capital assets must be adjusted for this depreciation.

Elementary depreciation methods

In elementary bookkeeping, it is customary to show the value of each fixed asset on the balance sheet at its opening book balance as at the beginning of the financial year less the deduction for depreciation as charged against profits, the final amount and up-to-date value of the asset in the balance sheet corresponding with the adjusted ledger balance of the asset.

The accounting entries are thus:

▶ *a* debit *to profit and loss account of the depreciation amount, reducing the trading profit, and*
▶ *a* credit *to the particular asset ledger account, with the consequent reduction of its book value.*

Apart from the annual and periodic re-valuation of a fixed asset (like loose tools) with the difference in value being adjusted against profits, the two main depreciation methods encountered at basic accounting level are known as the **fixed instalment** (or sometimes 'straight line') **method** and the **reducing instalment** (or reducing balance) **method**. The differences in approach of these two methods to determining the appropriate depreciation cost for the year is best explained by using an illustration.

Ten-year Lease – fixed instalments of £1000 a year

Year 1			£				£
Jan. 1	Cheque (original cost)		10000	Dec. 31	Depreciation P&L a/c		1000
				Dec. 31	Balance c/d		9000
Year 2							
Jan. 1	Balance b/d		9000	Dec. 31	Deprec. P&L a/c		1000
				Dec. 31	Balance c/d		8000
Year 3							
Jan. 1	Balance b/d		8000				

Van Account – 25% on the reducing balance

Year 1			£	Dec. 31	Deprec. P&L a/c		£
Jan. 1	Cheque (original cost)		3200				800
				Dec. 31	Balance c/d		2400
Year 2							
Jan. 1	Balance b/d		2400	Dec. 31	Deprec. P&L a/c		600
				Dec. 31	Balance c/d		1800
Year 3							
Jan. 1	Balance b/d		1800				

Note the difference between the two methods of elementary depreciation illustrated. The ten-year lease is depreciated by equal instalments of £1000 a year, whereas the van is depreciated by 25% *on the reducing balance of the account*, the depreciation instalments charged against profits therefore reducing each year.

The sections of the profit and loss accounts and the balance sheets applicable to the depreciation charges and the fixed asset account details will be shown thus:

Profit and Loss Account

Year 1			£	£			
	Deprec. of lease		1000				
	Deprec. of van		800	1800			
Year 2							
	Deprec. of lease		1000				
	Deprec. of van		600	1600			

Balance Sheet

Year 1	Fixed assets	£	£	£
	Lease	10 000		
	Less depreciation	1 000	9 000	
	Van	3 200		
	Less depreciation	800	2 400	11 400
Year 2	Fixed assets			
	Lease	9 000		
	Less depreciation	1 000	8 000	
	Van	2 400		
	Less depreciation	600	1 800	9 800

Scrap value

In instances where the fixed instalment of depreciation method is adopted for machinery or vehicles, the working life of the asset is first estimated, together with its likely scrap value on sale or disposal. The scrap value is then deducted from the original cost of the asset, and the remainder divided by its estimated number of working years, to give the annual charge against profits for *fixed instalment* depreciation.

Say a machine's original cost is £4000, and its working life estimated at 12 years, with a possible scrap value of £400 on its replacement. The fixed annual charge for depreciation would be (£4000 – £400)/12 = £300 a year.

For some assets the scrap value may, of course, be zero. For example, this is very likely to be the case for the 10-year lease in the example above. In such cases you can just use zero for the scrap value in the depreciation computation. For example, for the 10-year lease:

Depreciation = (£10 000 – 0)/10 = £1 000 a year (as the example showed)

Provision for depreciation account

It is a generally accepted accounting principle that fixed assets must be shown on the balance sheet at their cost price, the revenue account being debited in the normal way with the annual charge for depreciation with the corresponding credit being taken to a Provision for Depreciation Account, instead of taking it to the credit of the asset account directly (as we did above).

This method is more informative to all those interested in the true finances and net worth of a business as the original values of assets, and their respective accumulated depreciation charges, then remain clearly visible. The aggregate and cumulative depreciation is built up in the Provision for Depreciation Account, and deducted in total each year, from the original cost of the asset, shown on the balance sheet, in this way:

Machinery Account
(Depreciation by fixed instalments
of £500 a year via provision account)

Year 1			£				
Jan.1	Cheque for new machine	CB	5000				

Provision for Depreciation Account

Year 1			£	Year 1			£
Dec. 31	Balance c/d		500	Dec. 31	Profit & loss a/c		500
Year 2				Year 2			
Dec. 31	Balance c/d			Jan. 1	Balance b/d		500
			1000	Dec. 31	Profit & loss a/c		500
				Year 3			
				Jan. 1	Balance b/d		1000

The profit and loss account will be debited with the fixed amount of £500 a year (as also would have been the case using the previous method), and the asset will be shown on the balance sheet at its *original cost less aggregate depreciation*, thus:

Year 1	Fixed assets	£	£
	Machinery at cost	5000	
	Less aggregate deprec.	500	4500
Year 2	Fixed assets		
	Machinery at cost	5000	
	Less aggregate deprec.	1000	4000

PAUSE FOR THOUGHT

Explain what a wasting asset is: give one or two examples, and say how the depreciation should be dealt with in the accounting records.

▶ *What is the purpose of writing off depreciation? Is the value of the asset increased by these entries in the accounting records?*

▶ *Apart from normal wear and tear, give two further instances where merchandise could lose value over a period of time.*

▶ *Why does the depreciation account provide a more accurate picture of an asset's value on the balance sheet?*

KEY POINTS

▶ The loss in value of fixed assets is a business expense and chargeable against revenue. If not accounted for, the balance sheet does not present a true picture of the net worth of the assets.

▶ The writing off of depreciation as a business expense does not in any way affect the cash position of the business. It is purely a book entry or paper adjustment at this stage. Look back to the cash flow statement chapter (p. 161) where this was discussed if you need to.

▶ When a fixed asset such as a machine or a vehicle is sold, the sum realized on sale is debited to cash and *credited to the asset account*.

The depreciation charges that have been written off in the past are calculated to the date of the sale, and a profit or loss figure ascertained, as the asset must now be *eliminated from the books*.

The balance or difference left on the asset account is finally transferred to the profit and loss account.

TESTING YOURSELF

24.1 A machine is anticipated to have a working life of 15 years with a scrap value of one-tenth of its original cost at the end of that period. The original cost was £3000. What depreciation charge should be written off each year if the fixed instalments method is adopted?

24.2 A lease costs £8000 for the term of ten years. Show the asset account for the first two years under what you might consider an appropriate depreciation method.

Special machinery costing £4000 is bought on 1 January. Show the machinery account for the first two years, calculating depreciation on the fixed basis of 15% on the reducing balance of the asset.

24.3 Rory Macadam asks you to prepare, from this abbreviated form of trial balance, his final accounts as at 31 December, taking note of the adjustments for depreciation shown underneath.

	£	£
Capital a/c 1 Jan.		8000
Gross profit on trading		9300
Drawings for year	4500	
Debtors/creditors	1250	880
Carriage on sales	45	
Sundry expenses	85	
Machinery 1 Jan.	5400	
Additions during year	600	
Van account 1 Jan.	1200	
Fittings/equip. 1 Jan.	520	
Discounts	50	60
Wages owing 31 Dec.		300
Salaries for year	4200	
Stock 31 Dec.	2200	
Bank overdraft 31 Dec.		1510
	20050	20050

Allow for depreciation as follows:

a 10% on the original cost of the machinery which was £9000 *(Ignore additions, as these were bought in December.)*

b 20% on the ledger balance of the van account

c 5% in fittings/equipment account.

25

Bad debts and provision for bad debts

In this chapter you will learn:
- *the difference between bad and doubtful debts*
- *how to account for bad debts*
- *how to provide for doubtful debts*

The balances of those debtors who cannot or who persistently refuse to pay their accounts, may be eventually 'written off' to a **bad debts account** to remove them from the normally collectable debtors totals, although proceedings for the recovery of the money owing may continue for some time. Debts which are regarded as irrecoverable, like other business losses and expenses, are transferred at the end of the trading period, via the bad debts account, to the debit of the profit and loss account.

Later, should the debt or part of it be recovered, the sum received will be debited to the cash book and credited to a **bad debts recovered** account. Alternatively, the amount recovered from an old debtor may simply be taken to the credit of the current bad debts account, thereby reducing the *net* debit for the year.

Insight

Note, however, that amounts recovered from past defaulters are *not credited to their old personal accounts* in the sales ledger as these accounts will have been closed.

Provision for bad debts

Some debts are doubtful, not necessarily bad, though in time a proportion may prove to be irrecoverable. A *specific provision* for bad debts, covering those debts regarded as doubtful, is sometimes made on the lines now to be described. Note that this *provision is additional* to any debts known to be bad that are being written off as deemed irrecoverable.

The provision is normally a small percentage (perhaps 2 or 3%) of the total debtors' balances (net balances after deducting actual bad debts already written off).

In the first instance, when creating the new provision, say 3% of net debtors of £2000, profit and loss account would be debited with £60, and a credit for the same amount (£60) taken to a new account headed **provision for bad debts account**. The credit balance on this account is shown as a deduction from the trade debtors' total under current assets in the balance sheet.

Insight

It is important to bear in mind, though, that *this is a paper entry, simply a provision against a possible contingency only*, as all traders and businesses would like to see their doubtful accounts recovered in full.

Once created, the provision account remains a credit balance in the books, subject to annual adjustment, usually dependent upon the increases or decreases in the trade debtors' total at the end of the financial year. In the illustration now shown, note that after the first year (the creation of the provision) only the amount of the *provision increase is debited to revenue*; alternatively, any *provision decrease is credited back to revenue*.

	Debtors' listed totals at year-end £	Actual bad debts to be written off £	Bad debt provision
Year 1	2100	100	3% of net balances
Year 2	3150	150	3% of net balances
Year 3	2565	65	3% of net balances

The actual bad debts (in the centre section) will be transferred from the sales ledger accounts to the debit of the bad debts account, and then written off to the profit and loss account at the end of the trading year.

Provision for Bad Debts Account

Dr.   Cr.

		£			£
Year 1	Balance c/d	60	Year 1	P&L a/c (3% of £2000)	60
Year 2	Increased prov. carried down at end of second year c/d	90	Year 2	Balance b/d P&L a/c (provision increase to 3% of £3000)	60 30
Year 3	Decreased prov. to 3% of £2500 P&L a/c Balance c/d	15 75	Year 3	Balance b/d	90
			Year 4	Balance b/d	75

Profit and Loss Account

		£			
Year 1	Bad Debts a/c	100			
	Prov. for B/Ds a/c	60			
Year 2	Bad Debts a/c	150			
	Prov. for B/Ds a/c	30			
Year 3	Bad Debts a/c	65	*Year 3*	Prov. decrease written back	15

Balance Sheet

		£	£
Year 1	*Current assets*		
	Trade debtors	2000	
	Less provision	60	1940
Year 2	*Current assets*		
	Trade debtors	3000	
	Less provision	90	2910
Year 3	*Current assets*		
	Trade debtors	2500	
	Less provision	75	2425

Note that the actual bad debts each year are first deducted from the sales ledger balances *before* the operation of the provision. When first created, the new provision amount (£60) is debited to the profit and loss account and credited to the new provision account. The trading profit is reduced by this £60, and on the balance sheet the trade debtors are reduced by the same amount.

In the second year, the net debtors' total has increased to £3000. Again, maintaining the same provision percentage, the provision is now to be increased to £90 (3% of £3000). But there is already a credit of £60 on the provision account, so that it is only necessary to debit an additional sum of £30 to the profit and loss account in this second year. The corresponding credit is taken to the provision account, thereby increasing the provision to £90 at the end of this second year.

In the third year, the total of the trade debtors has been reduced to £2500, so that a provision of £75 (3% of £2500) only is required for this year. But since we already have a credit of £90 on the provision account, £15 of this credit balance will now be recouped by the reverse process of crediting the profit and loss account with the £15 written back, and debiting the provision account with £15. This now leaves the balance of £75 as the adjusted and correct new provision as at the end of the third year.

Note that it is always the final adjusted figure on the provision account which is to be deducted from the trade debtors' total shown under current assets on the balance sheet.

PAUSE FOR THOUGHT

▶ *What is the difference between writing off as a bad debt the account of a defaulting customer, and the creation of a specific provision for bad debts?*
▶ *Explain the difference between insolvency, being financially embarrassed temporarily, and being made a bankrupt.*

KEY POINTS

▶ Sometimes the provision account is combined with the bad debts account, the balance of the joint account being taken to the profit and loss account at the year end.
▶ The main difference between the two accounts, though, is that the bad debts account is a record of actual loss and expense suffered by the firm, whereas the provision account may be looked upon as more of a form of insurance against possible loss, and to show the trade debtors on the balance sheet at a conservative figure.

- ▶ Amounts recovered from past defaulters are credited to the current bad debts account, or sometimes to a bad debts recovered account. In either case, they serve to reduce the actual bad debts at the end of the trading session.
- ▶ After the creation of a new provision for bad debts, only the *difference between the old and the new provisions* is transferred to the profit and loss account, to the debit if an increase and to the credit if the provision is reduced.
- ▶ The net total of the trade debtors (with actual bad debts already deducted) is shown under current assets on the balance sheet, *less the last credit balance* shown on the recently made up provision for bad debts account.

TESTING YOURSELF

25.1 Luellen Blake has built up a small but profitable retail business, mainly on credit trade. Three years ago her accountant introduced an annual provision of 5% as a safeguard against bad and doubtful debts.

Her trade debtors' balances for last year totalled £4300 and for the current year £5800. The actual bad debts written off debtors' accounts and transferred to the bad debts account amounted to £32 this year, and a payment of £24 had been received on account of an old debt written off two years ago.

Make up Miss Blake's bad debts and her provision for bad debts account for the year just ended at 31 December.

25.2 On 30 June a year ago, the provision for bad debts in Matthew Cole's books showed a credit balance of £440. During the year he had written off actual bad debts of £72, and an amount of £18 was recovered from an old customer who had returned from Melbourne; his ledger account had been written off two years ago. Trade debtors' balances for the current year totalled £7400.

You are required to maintain the provision at 5%, make up the main accounts affected, and show how the profit and loss account of Mr Cole is adjusted.

25.3 On 31 December, the following sales ledger balances were extracted from the books of Cavendish Smith.

	£		£
Angela Ward	2800	Godfrey Bean	630
Trisha Benton	980	Rebecca Lawson	1580
Pamela Potter	860	Oliver Fairfax	1650

During the trading year ended 31 December two small accounts totalling £360 had been written off to bad debts account. The trustee in bankruptcy of one of these debtors on 20 December paid a dividend of 25% on Cavendish Smith's claim for £240.

The bad debt provision on the previous balance sheet was £480. It was decided to reduce this to 4% on the total debtors' balances for this year.

Make up the bad debts and provision for bad debts account, and show how the provision account will appear on this year's balance sheet.

26

Year-end adjustments

In this chapter you will learn:

- *how to adjust for prepayments*
- *how to accrue for expenses incurred*

The costs and expenses *incurred* during an accounting period are not necessarily those which are paid during that same period. The *real income*, too, earned during that period, may also be quite different from the actual payments received from customers and income from other sources.

Certain adjustments and amendments to the trading profit of a business have been explained in the last two chapters. Further adjustments are often, however, also necessary to arrive at the true expenditure and the correct trading profit. We will explore these in this chapter.

Prepayments or payments in advance

A payment made in advance indicates that money has been paid before the last date of the accounting period for a *benefit still to be received* (in a future accounting period). For example, this often applies to property rates, where advance payment is always demanded, and to various insurances paid on certain dates throughout the business year, normally providing cover for the full calendar year.

The adjustment is made on the nominal account, the amount not yet used up being credited to the account and brought down as a debit balance, and shown on the balance sheet under the heading of *current assets*. The revenue debit is reduced by this advance payment, thus increasing the trading profit by the same amount.

In the example now shown, the annual premiums are paid to the insurance company on 31 March and 30 June providing cover for the full twelve months. Assuming the financial year ends on 31 December, a proportionate adjustment must be made charging only £210 against the profits in this year, and £110 is carried forward for unexpired insurance, and shown as a current asset on the balance sheet.

Insurance Account

Dr.			£	Dec.			Cr.	£
Mar.								
31	Cheque	CB	200	31	P&L a/c			
June					¾ of £200 = £150			
30	Cheque	CB	120		½ of £120 = £60			210
				31	Prepayment c/d			110
			320					320
Jan.								
1	Balance b/d		110					

Accrued or outstanding expenses

Certain expenses may be owing at the end of the accounting period. Among these typical expenses are wages and salaries due to the firm's own employees, and accounts probably invoiced but unpaid such as for advertising, gas, electricity and printing.

Again, a two-way adjustment is involved, this time *increasing the debit to revenue*, and bringing down, on the actual expense account, the *amount owing as a credit balance*, to be shown under current liabilities on the balance sheet.

Weekly wages when paid in cash are normally paid on a Thursday or a Friday, made up to the preceding day or sometimes two days beforehand.

In the wages account shown below, the sum of £9040 has actually been paid out up to 28 June, but since the firm's financial year ends on 30 June, two days' pay (say £80) is owing to the daily paid employees at the balance sheet date.

The wages account is adjusted in this way:

Wages Account

Dr.			£	June			Cr. £
June							
30	Total wages actually paid (CB summary)		9040	30	Trading account (Wages paid plus wages due)		9120
30	Wages owing c/d		80				
			9120				9120
				July 1	Creditors for wages b/d		80

The credit balance on wages account will be shown on the balance sheet at 30 June, following trade creditors, under current liabilities.

ADJUSTMENTS IN FINAL ACCOUNTS: ILLUSTRATION

Robert Morris owns a small machine shop and retail business. The following balances were extracted from his books on 31 December, the end of his financial year.

Trial Balance 31 December

	£	£
Capital account: R. Morris 1 Jan.		30 000
Drawings for year	3 500	
Freehold premises at cost	25 000	
Machinery/equipment a/c bal. 1 Jan.	3 000	
Fittings, book balance 1 Jan.	400	
Van, book balance 1 Jan.	600	
Trade debtors/creditors	1 800	1 400
Bank overdraft		730
Cash in hand	25	
Bad debts provision account		120
Purchases and sales	8 500	23 250
Returns inwards/outwards	110	75
Carriage on purchases	145	
Carriage on sales	185	
Lighting/heating	330	
General expenses	665	
Discounts allowed/received	95	270
Commissions received		2 135
Advertising	400	
Bad debts written off	25	
Wages (storehouse and shop)	8 000	
Office salaries	4 000	
Stock on 1 January	1 200	
	57 980	57 980

In making up the final accounts of Robert Morris as at 31 December, the following adjustments and amendments are to be taken into account.

  a *Stock at 31 December valued at £1450.*
  b *Rates pre-paid £400 (in general expenses account).*
  c *Wages owing on 31 December £600; salaries owing £150.*
  d *The bad debts provision account stands at £120 in the books. The provision for the current year is to be reduced to 5% of the debtors' total at 31 December.*

e During the year Mrs Morris, wife of the proprietor, had taken goods for family use amounting to £245 at cost.

f Fittings are to be depreciated by 10% of their book value; vans by 20%, and machinery and equipment also by 20% of their book value at 1 January.

g Carry forward 50% of the advertising debit, as this was recently incurred, and the benefit expected in the New Year.

h The lighting/heating debit for £330 is to be split evenly between the trading account and the profit and loss account.

The model answer to this exercise should prove useful to the student for reference.

Trading and Profit and Loss Account for the year ended 31 December

	£	£	£
Sales		23 250	
less returns inwards		110	
			23 140
Cost of sales:			
Opening stock		1 200	
Purchases		8 500	
Carriage inwards		145	
		9 845	
Less returns outwards	75		
Goods for own use	245		
		320	
		9 525	
Wages (inc £600 owing)		8 600	
Light and heat (50%)		165	
		18 290	
Less closing stock		1 450	
Cost of sales			16 840
Gross profit			6 300

Discounts received			270
Commissions received			2135
Decrease in bad debt provision			30
			8735
Salaries (inc £150 owing)		4150	
Carriage outwards		185	
Light and heat (50%)		165	
Discounts allowed		95	
Bad debts written off		25	
Advertising (£400 less £200 c/f)		200	
General expenses (£665, less rates in advance £400)		265	
Depreciation:			
Machinery	600		
Van	120		
Fittings	40		
		760	
			5845
Net profit			2890

In many small retail businesses, some of the available stock for sale (in particular foodstuffs) is taken for the use and consumption of the proprietor's family. The correct adjustment is to *debit the owner's drawings account and credit the purchases account* at the cost price of the merchandise. In this instance the cost of the goods withdrawn by Mrs Morris (£245) is deducted from purchases on the debit of trading account, and added to her husband's drawings in the capital account section of the balance sheet.

An ordinary bank overdraft is a current liability, but a long-term loan made to the business (or a mortgage) is shown as a *fixed liability* under its own heading, positioned between the capital account and the current liabilities.

The new provision for bad debts is £90 (5% of the debtors' listed total). This new provision supersedes the old provision of £120 shown on the trial balance. This means that the £30 excess is no longer required and may now be **written back** (credited) to the profit and loss account. Remember, too, that it is always the *last provision* to be deducted from trade debtors on the balance sheet.

The reason for the carry-forward of £200 advertising has been explained in the list of adjustments. The nominal account for advertising will have been adjusted to show this £200 as a *debit balance*, but remember that the expense has already been paid.

Balance sheet

The balance sheet of Robert Morris as at 31 December, drawn up below, shows the net assets and the working capital of the business.

> **Insight**
>
> For a full set of accounts, comparison with last year's figures would be shown in brief in the right margin.

Balance Sheet of Robert Morris as at 31 December

Assets employed			
Fixed assets		£	£
Freehold premises at cost		25 000	
Fittings less depreciation		360	
Machinery/equipment less depreciation		2 400	
Motor van, less depreciation		480	28 240
Current assets		£	
Stock 31 Dec.		1 450	
Trade debtors	£1 800		
Less provision	90	1 710	
Rates in advance		400	
Cash in hand		25	3 585

Current liabilities			
Trade creditors	1 400		
Bank overdraft	730		
Expense creditors	750	2 880	
Net current assets			705
Net assets			28 945
Advertising expenditure carried forward			200
			29 145

Financed by

Proprietary capital

	£	
Capital account: Robert Morris 1 Jan.	30 000	
Add net trading profit for year	2 890	
	32 890	
Less drawings (£3 500 + £245)	3 745	
		29 145

KEY POINTS

▶ A prepayment (or payment in advance) refers to a benefit still to be used up. The amount paid in advance is deducted from its related debit on the profit and loss account, thereby increasing the trading profit.

The debit balance brought down on the expense account is listed on the balance sheet after trade debtors.

▶ Year-end adjustments are dealt with between the trial balance and final account stage. You may find it useful to mark the items affected by these adjustments with the pencilled amount of the adjustment to be made.

▶ An accrual or outstanding expense (wages/salaries owing and various expense bills unpaid) increases the debit to the trading or profit and loss account and reduces the net profit.

(Contd)

The credit balance brought down on the nominal expense account will be shown after trade creditors on the balance sheet.

In so far as *expense* creditors are concerned, think of the debt or money owing as being owed to a firm or organization which has already supplied your own firm with gas, water, electricity etc.

TESTING YOURSELF

26.1 This is the insurance account of a small trader at 31 December, the end of his trading year.

Insurance Account

Dr.			£				Cr.
Mar. 31	Fire insurance	CB	360				
Jul. 1	Motor insurance	CB	840				
Oct. 31	Employers' liability	CB	180				

All payments relate to full premiums paid. Complete this account to show the correct debit to revenue at the year end, and also show the balance sheet details.

26.2 Geoffrey Webb works a five-day week. Wages owing to his employees on 30 June amounted to £662. His wages sheets made up to the last pay day in the current year totalled £18 444. Show the net debit to revenue and how the item 'Wages owing' will appear on the balance sheet.

26.3 Valerie Seal balances her books on 31 December. Among other items on her trial balance are these two debits:

Electricity £180   :   Rates paid £260

The invoice for last quarter's electricity £38 was received early in the New Year. The amount of £180 refers to payments for the three previous quarters.

A cheque for £130 for the second instalment of the full year's rates had been paid on 8 October.

Miss Seal has been in business for one year. Complete her
nominal ledger accounts to show the net debits to revenue,
and also show how the adjusted items will appear on her balance
sheet at 31 December.

Examination exercise

Sole Trader's Accounts   First-year paper
ALL QUESTIONS TO BE ANSWERED   TWO HOURS ALLOWED

Insight

1 In bookkeeping examinations, accuracy and neatness are mark-earning features, and the ability to *work at speed* is equally important, as marks cannot be awarded for questions not attempted.

2 Establish your confidence by tackling the smaller and easier questions first, ensuring that they have in fact been answered to the best of your ability, and that you will not be penalized through running out of time in the rush work of in the last half-hour.

3 Accept the fact that few examinees manage to get *their accounts* to balance. A good deal of time is lost by candidates looking for mistakes made in balancing, time which could be put to better purpose. It is likely to be far more rewarding to aim at good average marks on the examination paper as a whole, rather than spend too much time on a large question, even though this question will carry maximum marks. Work as fast as you can on the larger problems, but when difficulties occur in balancing, and after a rapid re-check there is still a difference, *proceed to the next question* with all speed. If there is still time left before the papers are handed in, you can renew your re-checking after the completion of the remainder of the paper.

1 When checking and preparing information for the final accounts of Keith Warren for his financial year ended 31 December, you make a note of the following listed items.

a Repairs to van £68 had been charged to Van (asset) account.

b A payment of £25 from Marilyn Ross, sub-tenant of Keith Warren, has been debited to the business rates account.

c The trustee in bankruptcy of Sara Booth has recently sent a cheque for £36, being the first and final dividend of 40p in the pound of a sum written off her account two years ago. The firm's cashier has posted the payment to the credit of the debtor's old account.

d Included in purchases is an amount of £820, and in wages an amount of £350, both payments made in connection with the construction of a new showroom for the business.

e The sales ledger account of Hilton Mills has been credited with a cheque payment of £40 received from Milton Hills.

Make the necessary amendment or correction through the journal.

2 The warehouse stock of Sharon Lane was destroyed by fire in the early hours of 1 July, except for a small salvaged amount valued at £240. Miss Lane's accounting period was the normal calendar year. Her accountant provides you with this information.

	£
Stock valuation 31 December (six months previous)	2 200
Purchases (1 January to 30 June)	6 366
Net sales (1 January to 30 June)	14 880
Warehouse wages for six months	5 600

Prepare a draft trading account from this information, taking as the approximate figure for the gross profit of three-yearly average of 30% based upon the turnover of Miss Lane's business for the six months ended 30 June. State the amount of the claim for the destroyed stock to be made against the insurance company.

3 Titian Molloy owns a small dress-making business. The account balances listed below were extracted from her books at 30 June, the close of her financial accounting year.

For information only, the loan to Miss Molloy's friend Dinah Marsden is only of a temporary nature, to be repaid early in July. Interest has been waived.

Trial Balance 30 June

	£	£
Capital account 1July		12 000
Drawings for year	4 800	
Fittings/fixtures 1 July	1 420	
Additions during year	280	
Machinery/equipment 1 July	2 800	
Stock 1 July	1 450	
Purchases and returns	6 600	240
Sales and returns	280	18 680
Carriage inwards	88	
Carriage outwards	135	
Trade debtors/creditors	2 400	1 310
Expense creditors		75
Bad debts provision		160
Bad debts to be written off	120	
Wages of machinists	5 340	
Office salaries	3 800	
Discounts allowed/received	66	235
Commissions received		2 540
Advertising	1 350	
Lighting, heating and power	1 250	
Rent and rates	900	
Postages/stationery	36	
General expenses	72	
Cash at bank	1 023	
Cash in hand	30	
Loan to Dinah Marsden	1 000	
	35 240	35 240

In making up the final accounts for Miss Molloy, you are to take into account the following year-end adjustments.

- **a** Closing stock at 30 June was valued at £2850.
- **b** Insurance pre-paid £25 (debited to general expenses).
- **c** Lighting, heating and power expense is to be apportioned between the trading account and the profit and loss account three-fifths and two-fifths.
- **d** The advertising debit comprises a special campaign launched in May, and £500 is to be carried forward into the next financial year.
- **e** The bad debt provision is to be maintained at 5% of the total trade debtors' figure.
- **f** When questioned about goods withdrawn, Miss Molloy confirmed that she had taken material for her own use at the cost price of £180.
- **g** Fixtures and fittings are to be depreciated by 20% on the book balance of the asset, including the additions.
- **h** Machinery and equipment is to be depreciated by 20% on the book value at the beginning of the year.

4 a There is a credit balance of £120.56 in the bank column of your cash book. This has been brought down from the debit on 31 December. Would you show this on your balance sheet as a: **i** fixed asset **ii** current asset **iii** current liability?
  b If your bad debt provision is decreased, would this mean: **i** a smaller gross profit **ii** a larger net profit **iii** an increase in trade debtors on the balance sheet?

5 Post up the office equipment account of Anita d'Alvis from the information now given.

		£
1 January	Balance of account brought forward from last year	280
22 March	New filing cabinet bought	95
17 April	Old cabinet sold for (originally cost £35; shown in books at cost)	22
28 July	Bought new table and chairs	240

	Also timber for shelving	56
	Paid carpenter for work done	24
12 August	Sold old table and chairs (at original cost of £48 in books)	20
5 December	Paid for repairs to an old cupboard and desk	38

6 These are entries in the rent account of RHJ Textiles Ltd:

		£			£
Jan. 5	Cheque – quarter's rent on 15 Main Street	200	Jan. 6	Cheque – quarter's rent on 20 Main Street	200

Two young sisters, Sara and Simone, came to Preston from Avignon ten years ago and took up employment at the local textile mill. They rented a house in Main Street and one sister still lives there. Sara married a wealthy widower five years back, bought another house in Main Street, and when her husband died went to live in Blackpool. The Company now pays the rent on Sara's house for one of the managers. At what address does Simone live, number 15 or 20?

27

Incomplete records and single entry

In this chapter you will learn:

- *how to prepare accounts where full records are not kept*
- *how to estimate destroyed stock for insurance claims*

Single entry

The term **single entry** suggests there is no complete system of double entry. Whilst accountants and HM Revenue and Customs generally prefer to see double entry books, single entry is the only way to deal with incomplete records. Where records are incomplete, accountants must make use of all the available information they can find, calculating some of the figures by simple reasoning processes.

Where few records are kept

If accounts have not been prepared before, and there are neither records of business property nor of revenue income or expenditure, the visible existing assets should be verified, listed and valued. Then, checking backwards through a process of questioning and elimination, it should be possible to arrive at an approximate figure of the asset/liability position one year back.

To a large extent, the accountant has to rely upon the client's memory. Often the biggest complication is to agree with the proprietor the amount of personal drawings. Even if a rough cash book has been kept, rarely will the physical cash on hand, or in the till, agree with the 'book' balance according to the cash book. In many cases such a discrepancy is assumed to be 'drawings'.

Bear in mind that if a company's books and records are incomplete that it may have failed to comply with the Companies Acts. Incomplete accounting records may also cause significant problems if HM Revenue and Customs enquires into the trader's tax return.

Profit statement

The calculation of the profit or loss by the increased (or decreased) net worth method simply entails the listing and comparison of the known or approximated value of the assets and liabilities between the two periods, say of one year ago and today. The accounting equation $A = C + L$ is then brought into use, where A = assets, C = capital and L = liabilities.

Calculation	£
a *List the known assets and liabilities at the year-end to establish the closing capital of*	*4800*
b *Similarly, as far as possible, list the known and approximated assets and liabilities of one year ago to establish the commencing capital of*	*3500*
c *Add back the approximated figure for proprietor's drawings*	*2200*
d *Deduct any additional capital brought into the business*	*800*

Statement of Profit

	£
Increase of net assets (£4800 – 3500)	1300
Add back drawings	2200
	3500
Deduct additional capital brought in	800
Trading profit	£2700

The closing capital account, shown in its conventional form on the balance sheet, is a check on the arithmetic of this estimated profit.

Capital account	£	£
Balance 1 January (one year ago)	3500	
Additions during year	800	
Ascertained trading profit	2700	
	7000	
Less drawings for year	2200	4800

The existence of a bank account makes working on incomplete records much easier. A double entry basis can be built up from a break-down summary of the recorded items on the bank statements, together with up-to-date information of the fixed assets, stock valuation and lists of trade debtors and creditors.

EXAMINATION QUESTION

A typical examination question is now illustrated. A statement of affairs (balance sheet) of Adam Smith has been established as at the beginning of the trading period, one year ago. This is followed by a cash summary in total form, and there are a number of year-end adjustments to be taken into account when making up the final accounts for the year.

Cash and bank have been lumped together, quite usual in this style of question.

Statement of Affairs as at 1 January

	£		£
Capital account: Adam Smith		Premises at cost	4000
Balance 1 Jan.	6000	Fittings/equipment	850
Trade creditors	775	Stock	930
Expense creditors	25	Trade debtors	770
		Cash/bank	250
	6800		6800

Cash summary for the year

	£		£
Cash/bank balances b/f	250	Office salaries & NIC	1800
		Wages & NIC	2460
Cash sales total for year	6980	Cash purchases	326
		New counter	440
Cash/cheques from credit		Payments to suppliers	4350
customers	5150	Proprietor's drawings	1800
		Insurance	95
		Rates	120
		General expenses	89
		Balance c/d	900
	12 380		12 380
Balance b/d	900		

Adam Smith's current assets and liabilities at the year-end were:

stock £1130 : trade debtors £810 : trade creditors £660

The year-end adjustments to be taken into account are:

a *Wages due to sales assistants £42*
b *Rates pre-paid at 31 December £30*
c *Included in general expenses is a payment of £25 for an electricity bill shown as owing on the statement of affairs a year ago.*
d *Another bill for £32 is now due for electricity.*
e *Depreciate fittings and equipment by 10% on closing balance.*

Solution
First of all, the figures for credit purchases and credit sales are established on these lines by making up *total accounts* for both debtors and creditors.

Total Debtors

		£			£
Jan. 1	Balance b/f	770	Dec. 31	CB receipts	5150
Dec. 31	CREDIT SALES	5190	Dec. 31	Balance c/d	810
		5960			5960
Jan. 1	Balance b/d	810			

Total Creditors

		£			£
Dec. 31	CB payments	4350	Jan. 1	Balances b/f	775
Dec. 31	Balances c/d	660	Dec. 31	CREDIT PURCHASES	4235
		5010			5010
			Jan. 1	Balance b/d	660

Note that there are four basic elements to this type of problem: commencing balances, purchases or sales for the period, payments made, and the closing balances. Given any three of these elements, the fourth can quickly be found.

**Trading and Profit and Loss Account
for the year ended 31 December**

	£	£	£
Credit sales			5 190
Cash sales			6 980
Total sales			12 170
Cost of sales			
Credit purchases		4 235	
Cash purchases		326	
Wages	2 460		
Wages owing	42	2 502	
Stock 1 Jan.	930		
Stock 31 Dec.	1 130	(200)	
			(6 863)
Gross profit			5 307
Salaries		1 800	
Insurance		95	
Rates	120		
In advance	30	90	
General expenses		96	
(£89 + 32 − 25)			
Depreciation of fittings		129	
			2 210
Net profit			3 097

Balance Sheet of Adam Smith
as at 31 December

Fixed assets	£	£
Premises at cost		4000
Fittings/equipment	850	
additions	440	
	1290	
less deprec.	129	1161
		5161
Current assets		
Stock 31 Dec.	1130	
Debtors	810	
Rates in advance	30	
Cash/bank	900	2870
		8031
Capital and liabilities	£	£
Capital balance 1 Jan.	6000	
add net trading profit	3097	
	9097	
less drawings	(1800)	7297
Current liabilities		
Trade creditors	660	
Creditors for wages	42	
Expense creditors	32	734
		8031

Remember that all year-end adjustments are two-sided. Both debit and credit entries are involved. Revenue gains and losses are adjusted on the profit and loss account.

Claims for loss of stock

Sometimes insurance claims for loss or damage to stock (perhaps through fire or flood) have to be based upon past records and whatever figures are available. The following is a typical basic examination problem.

Peter West's warehouse stock was destroyed by fire on 30 June, except for a salvaged amount valued at £350. The date of the fire was exactly half way through his accounting period, and these figures were extracted from his account books and papers, kept in a fire-proof safe:

	£
Stock at 1 Jan.	1 500
Purchases (Jan.–June)	6 300
Wages for six months	3 200
Sales (Jan.–June)	12 400

A trading account was prepared from this information and a figure approximated for gross profit based upon 25% of Mr West's turnover for the current year, as this percentage had been fairly constant in recent years and was confirmed to be the average within the same trade.

Make up the insurance claim for Mr West.

Answer

Trading Account
for period 1 January to 30 June

	£		£
Stock 1 Jan.	1500	Sales (Jan.–June)	12 400
Purchases	6300	Estimated stock	
Wages	3200	at date of fire	
	11 000	(£14 100 – £12 400)	1 700
Gross profit			
(25% of £12 400)	3 100		
	14 100		14 100

The insurance claim would be:

Estimated stock at 30 June	£1700
Less value of stock salvaged	350
	£1350

PAUSE FOR THOUGHT

There is a similarity between some 'incomplete record' type of questions and the accounts of non-trading concerns (see the following chapter). In both instances it is often necessary to establish the final cash/bank balance for the balance sheet; often, too, forms of total accounts are needed to work out the purchases/sales in one case, and the correct subscriptions for the season in the other case.

In first-level Accounting examinations, the likelihood of getting one of these problems is fairly high.

TESTING YOURSELF

27.1 Zoë Patel's accountant lists her assets and liabilities at the beginning and at the end of her second trading year as follows.

	1 January	31 December
	£	£
Stock	850	920
Trade debtors	440	550
Trade creditors	310	380
Expense creditors	—	25
Fittings and fixtures	660	1375
Loan from sister	1000	800
Bank balance	230	150

The accountant also finds out that Zoë had withdrawn a regular £30 a week for her personal use, and had taken provisions from the business for family use averaging £6 a week. She had sold her private car for £600 and bought £500 of new fittings and equipment for the business out of the car proceeds. Towards the end of the year, she withdrew £1000 from the business to put down as a deposit for a new private car.

Calculate Miss Patel's profit for the year and show her balance sheet as at 31 December.

27.2 Saffron St Cloud is a part-time milliner. Her statement of affairs at 31 December is now shown.

	£		£	£
Capital 31 Dec.	6000	*Fixed assets*		
Current liabilities		Machinery	4400	
Trade creditors £635		Van	600	5000
Bank overdraft 290				
Wages owing 75	1000	*Current assets*		
		Stock	1025	
		Debtors	960	
		Cash	15	2000
	7000			7000

Her transactions for the following month of January are summarized in total form and listed below:

		£
1	Cash sales (cost price of goods £175)	280
2	Paid into bank	100
3	Cash purchases	20
4	Paid cheque for second-hand van	800
5	Received cheque for old van	550
6	Additional stock bought on credit	420
7	Sales to credit customers (cost price £640)	880
8	Sent cheques to suppliers	330
9	Cheques received from trade debtors	590
10	Paid expense creditors (wages owing – see BS)	75
11	Paid various expenses in cash	35
12	Proprietor's withdrawals in cash	40

You are required to draft a new balance sheet for Miss St Cloud as at 31 January, taking into account all the above transactions and finding her net trading profit for the month of January.

27.3 Ralph Edwards' statement of affairs, drawn up on 1 January, a year ago, is shown in brief detail.

	£		£
Capital (the balancing figure)	16 500	Premises at cost	10 000
Trade creditors	1 500	Fittings/equipment	2 100
		Stock 1 Jan.	3 300
		Trade debtors	2 400
		Bank/cash	200
	18 000		18 000

His accountant estimated his drawings for the year at £8300, after making a cash summary of totals extracted from a rough cash book and the bank statements of Mr Edwards.

	£		£
Balance b/f	200	Cash purchases	1 800
Cash takings	18 130	Wages and NIC	6 600
Payments from		New counter & shelving	4 400
credit customers	9 270	Rates and insurance	450
		Payments to suppliers	5 500
		General expenses	120
		Drawings	8 300
		Balance in hand	430
	27 600		27 600

On 31 December, the end of his financial year, the current assets and liabilities were:

Stock £4500   Debtors £3200   Creditors £1800

You are required to make up a revenue account and a balance sheet at 31 December for Mr Edwards, taking three adjustments into the accounts:

Rates in advance £80

Wages owing £140

Fittings and equipment to be revalued at £6000.

28

Non-trading concerns and club accounts

In this chapter you will learn:
- *the special features of accounts of non-trading organizations*
- *how to prepare income and expenditure accounts*

Most clubs, societies and associations are non-trading organizations. They derive their main source of income from subscriptions and donations, and, as a general rule, allocate their surplus funds to the purpose or the project for which they have been formed; alternatively their surplus revenue is used to improve or develop their activities and amenities for their members.

Receipts and payments account

Small clubs and societies have few assets other than a local bank account and probably some printed stationery. The bookkeeping records of this kind of organization are simple and uncomplicated. The form of account produced by the treasurer at the end of the year or the season is easily understood by members. It is necessary only to draw up a receipts and payments account, which is simply a cash summary for the period covered by the account. There is an illustration of a receipts and payments account on the next page.

The receipts and payments account, shown below, is a brief summary of money received and money paid out during the year. Every receipt, subscription and donation is included in total summary, and every payment made is also included in total form. No distinction is made between capital and revenue expenditure; this of course would probably be unnecessary in the case of a small society as all incoming cash and payments would be regarded for revenue purposes only, with no attempt made to show financial gain or loss beyond the difference in the commencing and closing cash balances.

This statement is simply a summarized copy of the club's cash book, and cash and bank have been combined to make it easier for members to understand.

The Over-Sixty Social Club

Receipts and Payments Account for year ended 30 June

	£		£
Cash in hand 1 July	24.50	Postages/stationery	12.20
Subscriptions rec'd		Hire of local hall	150.00
(124 members at £2)	248.00	Lighting/heating	55.40
Donations	15.00	Refreshments	
		bought	102.80
Gross proceeds from		Prizes purchased	250.00
whist drives and		Donations to Oxfam	25.00
socials	350.00	General expenses	16.70
		Balance of cash in	
		hand 30 June c/d	25.40
	637.50		637.50
Cash in hand 30 June	25.40		

Income and expenditure account

The receipts and payments account is incomplete in so far as real financial information is concerned. Clubs and societies of any size acquire certain assets and property, and a cash summary would disclose very little about the true finances and net worth of the organization.

These larger clubs, societies and associations, at their annual general meetings, present to their members an income and expenditure account together with a balance sheet on the lines we already know. The long-sounding term 'income and expenditure account' is only another name for the profit and loss account with which we are familiar. Profit and loss would not, of course, be appropriate terminology for a non-profit oriented organization. Similarly the 'capital' becomes the 'accumulated fund' or occasionally 'general reserve'. Clubs and societies may also, for their own convenience, split their reserves into other categories; for example, a 'maintenance reserve' into which sums are transferred each year to meet periodic maintenance bills such as re-decorating.

Often an examination question presents the candidate with a receipts and payments account (or simply a summarized cash book, which is the same thing) and, subject to a number of adjustments, requires an income and expenditure account to be made up together with a balance sheet.

An illustration of a typical examination problem is now given.

The Heathside Tennis Club
Receipts and Payments Account for the year ended 31 October

	£		£
Balance 1 Nov.	97.20	Postages/stationery	37.80
Subs. rec'd (including £30 in advance)	920.00	Refreshments bought	124.40
		Dance expenses	70.40
Dance proceeds (gross)	137.00	New nets	460.00
Donations	50.00	General expenses	25.20

(Contd)

Receipts from teas and socials		244.60	Wages of groundsman		612.00
			Heating/lighting, etc.		83.40
			Balance c/d		35.60
		1448.80			1448.80
Balance b/d		35.60			

In making up the annual accounts as at 31 October for presentation to club members, the following information is to be taken into account:

The club's assets and property on 1 November, one year ago:

Clubhouse at original cost	£3000.00
Nets/equipment at book value	260.00
Stock of tinned drinks, etc.	43.00
Cash on hand (see cash summary)	97.20

There were no arrears of subscriptions one year ago, and all expense creditors had been paid.

Certain adjustments are now to be taken into account in making up the income and expenditure account and the balance sheet:

   a *invoice for electricity £22.80 unpaid at 31 October*
   b *subscriptions in arrears at 31 October amounted to £50*
   c *subscriptions in advance £30 (shown in cash summary)*
   d *wages owing to groundsman at 31 October £24*
   e *stock of refreshments at end of current season £49.60*
   f *nets/equipment account to be depreciated by £200.*

The first stage of the answer to this problem is to find the opening capital (sometimes called the accumulated fund), and then make up the subscriptions account to find the true income from members for the year.

Accumulated fund $\left\{\begin{array}{l}\end{array}\right.$	clubhouse	£3000.00	
	nets/equipment	260.00	
	stock 1 Nov.	43.00	
	cash balance 1 Nov.	97.20	£3400.20

Naturally, any liabilities at the beginning of the year would have been deducted from this amount of £3400.20, but in this instance there were no liabilities.

Subscriptions Account

Dr.						Cr.
Oct.		£	Oct.			£
31	Subs. paid in advance by members c/d		31	Actual receipts as per cash summary		920.00
		30.00				
31	Transfer to Income & Expd. a/c	940.00	31	Subs. owing by members c/d		50.00
		970.00				970.00
Nov.			Nov.			
1	Debtors for subs. b/d	50.00	1	Creditors for subs. b/d		30.00

Subscriptions taken to the income and expenditure account must show the actual income from members for the year.

The final accounts of the club can now be made up on these lines:

Income and Expenditure Statement – year ended 31 October

Income	£	£	£
Subscriptions		940.00	
Donations		50.00	
Gross receipts from dances, teas, etc.	381.60		
Adjustment on stocks	6.60		
	388.20		
Less dance & catering expenses	(194.80)	193.40	1183.40
Expenditure			
Postages/stationery		37.80	
Groundsman's wages	612.00		
Wages owing at 31 Oct.	24.00	636.00	
Heating/lighting	83.40		
Electricity owing	22.80	106.20	
General expenses		25.20	
		805.20	
Deprec. nets/equip. a/c		200.00	1005.20
Surplus to accumulated fund			£178.20

Balance Sheet of The Heathside Tennis Club as at 31 October

Fixed assets	£	£	£
Clubhouse at cost		3000.00	
Nets/equipment	260.00		
Additions	460.00		
	720.00		
Depreciation	(200.00)	520.00	3520.00
Current assets			
Stock of refreshments	49.60		
Debtors for subs.	50.00		
Cash in hand	35.60	135.20	
Current liabilities			
Subscriptions in advance	30.00		
Creditors for wages	24.00		
Creditors for electricity	22.80	(76.80)	
Net current assets			58.40
Total assets			3578.40
Capital			
Accumulated fund 1 Nov		3400.20	
Add surplus income		178.20	3578.40

KEY POINTS

▶ In all problems involving income and expenditure, first
establish the closing cash balance needed for the final
balance sheet. Also make up a subscriptions account, or
work out the subscriptions by arithmetic.

(Contd)

- ▶ Sometimes a club has a small bar and/or offers catering facilities to members. A short trading account is then made up in the ordinary way, and the profit or loss on trading is carried down to a general income and expenditure account.
- ▶ Do not confuse subscriptions in advance with rates in advance. If club members have paid subscriptions in advance, they become creditors of the club; if their subscriptions are in arrears, they are debtors of the club.

TESTING YOURSELF

28.1 Make up a receipts and payments account from this information listed by the treasurer of a small local dramatic society at 31 December.

	£
Opening cash/bank balances (combined)	115.50
Subscriptions actually received from members during the year	150.00
Donations received	25.00
Net receipts from rehearsals/socials etc.	14.80
Ticket sales on productions	280.60
Rent of local hall	130.00
Stamps/stationery	6.20
Purchase of reading material and plays	84.50
Payment of royalties	62.00
Lighting/heating	28.70
Refreshments and drinks bought	92.40
Honorarium to treasurer 31 December	10.00

28.2 The cash book of the White Collar Workers' Association was analysed in total as follows: subscriptions from members £860; donations received £430; sundry receipts £175; rates paid £120; furniture bought £270; salaries paid £600; fees to lecturers £250; general expenses £210. Cash balance 1 January £30.

The Association leased its premises with the leasehold shown at cost of £4000. The old furniture had been given to the Red Cross and written off the previous year. The position with regard to outstandings at the year-end was as follows:

Subs. owing at 1 Jan. £46; and at 31 Dec. £66. Creditors for stationery £76 at the beginning of the year, and £24 at the end.

Take the valuation of stationery unused at £54, and make up an income and expenditure account and a balance sheet at 31 December. A payment of £130 for stationery by the association was found by the auditors to be debited to general expenses.

29

....

Control accounts

In this chapter you will learn:
- *the meaning of 'control accounts'*
- *how to prepare them*
- *why they are useful*

Self-balancing ledgers

A control account is used as a check on the accuracy of postings to a particular ledger. Before computers were available to make such calculations and checks almost instantaneously, this was a very important safeguard.

The term **self-balancing** refers to the inclusion of a control account, kept normally at the front or at the back of a ledger, and used as a check by the accounting staff on the total of balances extracted from that ledger.

In addition to providing a check on the work of the bookkeeper, self-balancing is also an aid to any internal audit system in force.

The bank cash book

The control accounts themselves are merely *summaries of totals taken from other books*, in the main from the books of original entry.

The rulings in these books of prime entry are adapted to suit the needs of the particular business and the control system adopted.

The bank cash book is a feature of the control system. The bank and discount columns are still used, but the old cash columns are replaced by 'details' columns. *All receipts*, cash and cheques, are *banked daily*, and all payments shown in the main cash book are *made by cheque*, automated transfer, direct debit or standing order, leaving the small routine cash disbursements to be paid by the petty cashier. The total wages paid each week will sometimes still be met by a cheque cashed for that purpose, but more usually along with the salaried employees they will be paid monthly by credit transfer.

The debit and credit details columns are extended in analysis form to provide the information required by the accountancy staff for their control summaries. For example, the cumulative total of the payments from debtors provides control with the total amount credited to all the customers' accounts in the sales ledger. Similarly all payments to suppliers are listed in one particular column on the credit side of the cash book, to establish the total amount paid to trade creditors during the period.

The general principles of control are best explained by a simplified illustration. In this example, we are concerned only with transactions on the sales side of the business, the credit sales to customers during one month, their payments on account of goods they have bought, discounts allowed to them and credits for goods returned. The purpose of the control account is to prove the arithmetical accuracy of the total of the debtors' balances extracted by the sales ledger clerk on 31 July.

ILLUSTRATION

The ledger accounts of only four credit customers are given for the month of July.

Smith

July		£	July		£
1	Balance	100	8	Cash	95
10	SDB	120		disct.	5
			12	SRB	15

Jones

July		£	July		£
1	Balance	50	18	Cash	48
20	SDB	75		disct.	2
			22	SRB	5

Robinson

July		£	July		£
1	Balance	20	25	Cash	19
				disct.	1

Brown

July		£	July		£
22	SDB	10	24	Cash	10
31	SDB	3			

The bank cash book (debit or receipts side only) is shown in brief detail, followed by the other books of original entry on the sales side, the sales day book and the returns inwards book.

		Fol.	Disct.	Sales ledger	Cash sales	Sundries	Bank
			£	£	£	£	£
July							
1	Balance b/f					400	400
8	Smith	SL	5	95			95
12	Cash sales	NL			80		80
18	Jones	SL	2	48			
	Cash sales	NL			30		78
24	Brown	SL		10			
25	Robinson	SL	1	19			29
31	Cash sales	NL			40		40
			8	172	150	400	722

Sales Day Book

July			£
10	Smith	SL	120
20	Jones	SL	75
22	Brown	SL	10
31	Brown	SL	3
			208

Sales Returns Book

July			£
12	Smith	SL	15
22	Jones	SL	5
			20

Again it is emphasized that this small illustration concerns only sales control, in connection with the items and transactions affecting the sales or debtors' ledger only.

At the month end, the credit customers' balances are extracted and listed as follows:

	£	
Smith	*105*	
Jones	*70*	
Brown	*3*	
to give the total	*£178*	*owing by debtors*

The accuracy of the postings to the sales ledger is then proved by making up the control account, either at the front or at the back of the sales ledger. This account contains summaries and totals, the figures being extracted from the bank cash book, the sales day book and the returns inwards book. The control account is called the general ledger control account, made up on these lines:

General Ledger Control Account for July

July			£	July			£
31	Sales returns	SRB	20	1	Balance b/f		170
31	Total payments			31	Total sales for		
	received	CB	172		month of July	SDB	208
	& discount allowed	CB	8				
31	Balance c/d		178				
			378				378
				August			
				1	Balance b/d		178

The opening *credit* balance of the above control account (£170) is the sum total of the debtors' balances at the beginning of the month, 1 July. Note that the totals shown within this control account are on the reverse side to the individual items that are already posted to this ledger. There is always a *credit* balance on the control account kept in the sales ledger, generally referred to as the general ledger control account (in effect a contra account to the sales ledger control account kept in the general ledger).

The purchase ledger control follows a similar procedure to that briefly outlined on the sales side, except that again, of course, it will be in reverse of the sales procedure as already illustrated.

Insight

The main difficulty experienced by students is knowing either the heading of the control account or the correct side for their sales and purchases debits and credits, etc.

First make sure of the ledger in which the account is to be written up.

If it is the sales ledger, all totals will be in reverse (on the control account) to the actual items shown on the individual debtors' accounts.

If it is the purchases ledger, again reverse the items on the control account, to the various postings within the bought ledger.

KEY POINTS

▶ Note any mention of bad debts refers to the sales ledger. Bad debts are always shown on the same side as debtors' payments and discount allowed.

(Contd)

- ▶ Note also that small debit balances in BL and small credit balances in SL at the end of month must be shown above and below the totals when making up control accounts.
- ▶ Be careful with transfers between ledgers, increasing the debit in one ledger and decreasing the credit in another – and vice versa.

TESTING YOURSELF

29.1 Prepare the purchase ledger control account (as drawn up by the general ledger clerk) for the month of January and carry down the balance.

Jan.		£
1	Credit balances	30 884
	Debit balances	122
31	Purchases for month	41 350
	Paid to suppliers (cheques)	26 440
	Discounts received	586
	Small credit balances paid out of petty cash	45
	Returns outward	958
	Contras settling small debits	82
	Debit balances at month end	40

29.2 From the balances below you are required to make up the purchase and the sales ledger control accounts as they would appear at the back of their respective ledgers:

June		£
1	Bought ledger balances	2 433
	Sales ledger balances	3 486
30	Purchases for month	9 295
	Sales for month	13 840
	Paid to suppliers	7 616
	Discounts received	190
	Returns inwards	168
	Received from customers	12 327
	Discounts allowed	255
	Returns outward	177
	Bad debts written off	36
	Bills receivable	585
	Bills payable	480
	Dishonoured cheque of customer	42
	Debit balances transferred from sales ledger to bought ledger	58

30

Partnership accounts

In this chapter you will learn:
* *how to prepare partnership accounts*
* *how to split the profit in an appropriation account*
* *about interest on capital and partners' salaries*

The ordinary bookkeeping records of a partnership are kept exactly on the same lines as those of a sole trader. The cash book, the day books, and the customers' and suppliers' ledgers are posted up in the same way as that already shown. The net profit of the business is calculated in the same way as before, but, because there are now two or more owners of the business, their shares of profit have to be worked out separately.

The appropriation account

Since the trading profit is to be divided, not necessarily equally, the profit and loss section of the main revenue account is extended by an **appropriation account**. The trading profit for the period is split in the appropriation account, and divided according to the terms laid down by the partnership agreement or contract.

In the absence of any written agreement, or by implied
custom of the past, partners are presumed to share profits
and losses equally, irrespective of the size of their capital
accounts or their holding of partnership assets. This is laid
down in the Partnership Act of 1890.

Separate capital accounts

Partners' capital accounts have no direct bearing on their
profit-sharing ratios, unless there are specific terms laid down to
the contrary. The senior partner may have a capital of £8000 and
the junior a capital of £2000 and yet their written agreement
or past custom may indicate that their profit sharing ratios are
three-fifths and two-fifths.

Interest on capital

Interest is sometimes allowed as a form of compensation for those
partners with larger financial holdings in the firm. This interest is
the first charge made in the appropriate account against the net
trading profit before its division between the partners, the double
entry being a debit to the appropriation account and a credit to the
individual current accounts of the partners.

Current accounts

This is a new term referring to the private accounts of partners
within the firm, not to be confused with the ordinary bank current
account. A partner's current account replaces the old drawings
account of the sole trader. The total drawings of the individual
partner is debited to that partner's current account; the share of

profit and also any interest on capital or any partnership salary to which the partner is entitled are credited to the account.

It is customary to keep fixed capital accounts of all partners quite separate and distinct from the individual current accounts of each partner although the two of course need to be applied together to determine the full extent of the partner's investment in the partnership.

ILLUSTRATION

Ned and Fred are in partnership with capitals of £4000 and £2000 respectively. Their partnership agreement states that the net trading profits will be shared in the proportions of three-quarters and one-quarter, but that Fred should be credited with a partnership salary of £1000 a year before the final allocation of net trading profit. In addition, interest on capital of 10% per annum is to be credited to the current account of each partner. The net trading profit for the year under review, ended 31 December, was £13 600.

The credit balances of the partners' current accounts brought forward were: Ned £120 and Fred £80; the partners' drawings for the year were: Ned £9350 and Fred £4220. At the balance sheet date Fred had not yet withdrawn any part of his salary for the current year.

Show the appropriation account of the firm, the partners' capital and current accounts (as they would appear in the private ledger), and draft the capital and liabilities side of the balance sheet as at 31 December.

Profit and Loss Appropriation Account

		£				£
Interest on capital			Net trading			
Ned		400	profit b/d			13 600
Fred		200				
Salary of Fred		1 000				
Shares of profit						
Ned		9 000				
Fred		3 000				
		13 600				13 600

Capital Accounts (fixed original balances)

Dr.									Cr.
			Ned	Fred				Ned	Fred
					Jan.1	b/f		£4000	£2000

Current Accounts

Dr.								Cr.	
		Ned	Fred					Ned	Fred
Dec.		£	£	Jan.				£	£
31	Drawings	9350	4220	1	Balances b/f			120	80
31	Balances c/d	170	60	Dec.					
				31	Int. on capital			400	200
					Salary				1000
					Shares of profit			9000	3000
		9520	4280					9520	4280
				Jan.					
				1	Balances b/d			170	60

**Balance Sheet of Ned and Fred
as at 31 December**

Capital and liabilities section		
Capital accounts	£	£
Ned	4000	
Fred	2000	6000
Current accounts		
Ned: Balance 1 Jan.	120	
Int. on capital	400	
Share of profit	9000	
	9520	
Less drawings	(9350)	170
Fred: Balance 1 Jan.	80	
Int. on capital	200	
Salary	1000	
Share of profit	3000	
	4280	
Less drawings	(4220)	60
Current liabilities		—

Partners' salary

In the illustration just shown Fred has been credited with his full salary of £1000 (on his current account) *because it has not yet been paid*. He becomes a creditor of his own firm for the time being. When the salary is withdrawn, cash will be credited and his current account debited. If Fred had withdrawn part of his salary (on account) prior to the balance sheet date, the double entry would have been a credit to cash and a debit to partnership salaries account, with *no credit appearing on Fred's current account for that part of the salary already withdrawn.*

Amalgamation of sole traders

Occasionally, two or more sole traders decide to amalgamate their respective businesses and become partners. The bookkeeping entries in connection with the presentation of the balance sheet of the new firm are simple enough providing the capital account of *each sole trader is adjusted,* according to the terms agreed upon by the prospective partners, *before any attempt is made to draft a final balance sheet for the new firm.*

ILLUSTRATION

Ellen and Emily are cousins with small retail drapery shops, both rented, in the town centre of Hereford. They decide to become partners and save a considerable sum on competitive advertising and also on bulk buying. The date of their amalgamation is to be 1 April. Their last individual balance sheets are now shown at 31 March in the older 'T' style, to make the calculations clearer.

Ellen's Balance Sheet as at 31 March

	£		£
Capital	5800	Fittings/fixtures	1500
Creditors	450	Van at cost	1250
Bank overdraft	250	Stock of goods	2200
		Trade debtors	1550
	6500		6500

Emily's Balance Sheet as at 31 March

	£		£
Capital	6460	Furniture/fittings	750
Trade creditors	840	Tools/equipment	1320
		Stock	1840
		Trade debtors	2880
		Bank	510
	7300		7300

The partnership agreement of the new firm states that profits and losses are to be shared equally as from 1 April, and that certain amendments and adjustments are to be made to some of the asset values as shown on the two balance sheets dated 31 March, namely:

All furniture, fittings, etc., are to be reduced by 20% of their book value; the van is to be revalued at £800, and tools/equipment at £1000. Provisions are to be made against possible bad debts of 10% of the trade debtors' balances. Miss Ellen is to repay her bank overdraft privately, and the bank account of Miss Emily is to be taken over by the new firm. The stock on hand shown in each balance sheet is to be depreciated by 25%.

Answer
First of all re-draft the individual balance sheets of the sole traders at 31 March, subject to the partnership terms agreed upon.

		Ellen £	Emily £			Ellen £	Emily £
Capital accounts		4595	5242	Fittings/fixtures		1200	600
Trade creditors		450	840	Van		800	
				Tools/equipment			1000
				Stock		1650	1380
				Debtors *less* prov. for bad debts		1395	2592
				Bank			510
		5045	6082			5045	6082

The balance sheet for the new firm as at 1 April can now quickly be drafted by lumping together the figures shown above.

Balance Sheet of Ellen and Emily as at 1 April

Assets employed

Fixed assets

	£	£
Fittings/fixtures	1800	
Tools/equipment	1000	
Van at valuation	800	3600

Current assets

Stock on hand		3030	
Trade debtors	4430		
Less provision	(443)	3987	
Bank		510	
		7527	

Current liabilities

Trade creditors		(1290)	
Net current assets			6237
Total net assets			9837

Capital accounts

Ellen		4595
Emily		5242
		9837

Overdrawn current account

A partner who overdraws on the current account becomes a debtor to the firm. The current account shows, temporarily, a debit balance, and at the balance sheet date, must be switched over to the bottom of the listed assets. However, it would be ignored in calculating the working capital of the business.

Interest on drawings

Interest is sometimes charged on partners' drawings with a view to curbing heavy and frequent withdrawals on the firm's bank account. This interest on drawings is debited to the partners' individual current accounts in addition to the sums withdrawn, and is credited to the firm's appropriation account.

For example, if the senior partner of a professional firm, by agreement, withdraws £2000 at the end of each quarter, subject to an interest charge of 8% per annum, the individual debits to her current account for the year ended 31 December would be calculated in this manner:

		£
31 March	(interest 1 April–31 Dec.) 9 months at 8%	= 120
30 June	(interest 1 July–31 Dec.) 6 months at 8%	= 80
30 Sept.	(interest 1 Oct.–31 Dec.) 3 months at 8%	= 40
31 Dec.	(last day of trading year – no charge)	
		£240

This total of £240 would be credited to the firm's appropriation account at the end of the trading year, and the amount of £8240 (£8000 drawings plus £240 interest) would be debited to this particular partner's current account.

Goodwill in accounts

Goodwill has been defined as 'the likelihood that the old customers will continue to deal with the old firm' and as 'the benefit arising from connection and reputation'.

Occasionally there appears at the top of the fixed assets on a balance sheet the descriptive term 'Goodwill £___'. It looks impressive, particularly if it is a large amount, but actually it does not mean very much beyond the fact that a debit balance for this intangible asset is still being carried forward in the books. This debit balance is a paper entry from the past (probably created when the business last changed hands or when a new partner was admitted to the old firm). It is an intangible asset; you cannot see or touch it, and its value is indeterminable until the business is sold or there is a change in the ownership. Even on the sale of the business, the value of goodwill, as part of the purchase price asked by the vendor, is always problematical and subject to much discussion and deliberation before the final price is agreed upon by the new owner. In more advanced accounting exercises goodwill is written off over a period of time in much the same way as tangible assets are depreciated (see Chapter 24). This provision is called amortization.

Insight

You might like to check out *Get to Grips with Book Keeping* for more details on Goodwill and the more technical aspects of bookkeeping and record keeping more generally.

A new business has no goodwill. It has to be earned and developed. As the business expands, its connection and reputation increases, so that when it is sold or a change of ownership takes place the valuation of the business 'as a going concern' is worth substantially more than the listed value of its net assets and property. *The difference in value between the net assets and the actual purchase price is the goodwill of the business.*

ILLUSTRATION

Sam Brown has rented a corner shop in the suburbs for the past 20 years. He is now retiring, and his last balance sheet, dated 30 June, is shown below on the left. It has been checked over by George Smith, who agrees the vendor's price of £18 000 for the business, advertised with 'all assets and property to be taken over at balance sheet figures, and the trade creditors to be settled by the new owner'.

Balance Sheet of Sam Brown, vendor, as at 30 June

Assets and property		
Fixed assets	£	£
Van at valuation	1 950	
Fittings/equipment	7 800	9 750
Current assets		
Stock at 30 June	2 800	
Trade debtors	450	3 250
		13 000
Financed by		
Capital: S. Brown		
Balance 1 July	14 000	
Net profit for year	5 500	
	19 500	
Less drawings	7 500	12 000
Current liabilities		
Trade creditors		1 000
		13 000

Balance Sheet of George Smith, purchaser, as at 1 July

Assets and property

	£	£
Fixed assets		
Goodwill at cost	6 000	
Van at valuation	1 950	
Fittings/equipment	7 800	15 750
Current assets		
Stock at 1 July	2 800	
Trade debtors	450	3 250
		19 000
Financed by		
Capital: G. Smith		
Balance 1 July		18 000
Current liabilities		
Trade creditors		1 000
		19 000

Comparing the two balance sheets it will be seen that George Smith has acquired the whole of the assets and liabilities of Sam Brown at their valuation at 30 June, and, in the agreed purchase price of £18 000 has, in effect, paid a premium of £6000 in acquiring these assets. The capital account of the new owner thus becomes £18 000, the price he has paid for his new business. The excess amount paid over and above the valuation of net assets taken over (i.e. £6000) is debited to the goodwill account, and on his first balance sheet dated 1 July, is taken to the top of the fixed assets. This intangible asset, though only a paper entry, is a debit balance in the accounting records of George Smith, and will continue to be shown at the top of his assets on future balance sheets less amortization.

Sometimes, too, where a new partner is being admitted to an existing firm, the old partner(s) create a goodwill account, perhaps only temporarily, by debiting an agreed figure, say £10 000, to a new asset account headed goodwill, crediting amounts to their capital (or current) accounts in the proportions they previously shared profits and losses, shown in this way by journal entry:

Dec.		£	£
31	Goodwill account	10 000	
	Capital accounts:		
	A		6 000
	B		4 000
Goodwill account created on admission			
of C, credited to capital accounts of			
A and B in their old profit sharing ratios.			

The creation of a goodwill account provides some compensation for the past work and effort of the old partners. It often relieves the new incoming partner from finding a fairly substantial sum on admission to the firm, although generally he would still be expected to bring in a small introductory capital. The creation of goodwill in the accounts of a partnership can cause capital gains tax problems, so it is normally written off again immediately, rather than being carried forward.

KEY POINTS

▶ In the absence of written agreement or laid-down policy or custom, Section 24 of the Partnership Act of 1890 states that partnership profits and losses will be shared equally.
▶ Appropriations and charges against trading profit such as interest on capital and partnership salaries are debited to the appropriation account before final distribution of profit.
▶ When two or more sole traders amalgamate to become partners, revalue and adjust their capitals separately before

making any attempt to lump together the joint assets/liabilities of the new firm.

▶ Partnership salaries are debited *in full* against the firm's trading profit. If part of salary is not paid, it will be shown as a credit on individual partner's current account.

TESTING YOURSELF

30.1 Make up the appropriation account and the partner's current accounts from the information given below.

	Capitals £	Current a/cs £	Drawings £
Melon	4000	250 Cr.	6500
Lemon	2000	100 Dr.	6800

Lemon is to be credited half-yearly with a salary of £500. Each partner is to be credited with 8% per annum interest on capital. The net trading profit for six months ended 30 June was £14 500.

30.2 A, B and C share profits in the ratios of 3:2:1. Net trading for this year is £12 000 after charging C's salary in full. Interest on capital is allowed at 10%. Make up the partners' current accounts on the balance sheet at 31 December.

Cash Book

Jan.			£	June			£
1	Capital a/cs			30	½ yearly		
	A		5000		salary of C		500
	B		4000	Dec.			
	C		3000	31	Drawings		
					A		4800
					B		3500
					C		2500

30.3 Sue Pugh's assets and liabilities at 30 June are listed:

fittings £8000; bank overdraft £550; stock £1800; trade creditors £1750; trade debtors £1200.

She takes her friend May Day into partnership as from 1 July. Miss Day pays the sum of £2000 into the firm's bank account, and also brings in her own car (valued at £1500 and some useful equipment valued at £200).

The partnership agreement provides for interest on capital at 6% per annum, and the profit sharing ratios of the new firm are to be:

  Miss Pugh, two-thirds.
  Miss Day, one-third.

You are required to make up the balance sheet of the partnership as at 1 July.

Revision exercise 5

1 The Overlanders' Club was established ten years ago for social and recreational purposes, in particular travel abroad. The secretary of the club prepared the following summary of the cash book transactions for the year ended 31 March.

	£		£
Balance (cash and bank combined)	8040	Furniture bought	1200
Subscriptions received	9150	Catering expenses	2660
Donations received	280	Repairs/maintenance	150
Receipts from socials, raffles, etc.	3840	Rent of rooms	1800
		Wages of temporary staff	6200
		Lighting/heating	880
		Books/magazines	760
		Printing and stationery	220

The club rents two large rooms overlooking the park at the economic annual rent of £2400, paid quarterly, the landlord paying the rates. Fixtures and fittings on the last balance sheet stood at £6800. The committee decided to revalue this at £6500, including the additions bought earlier in the year.

Books and magazines were regarded as expendable, to be treated purely as revenue.

The position with regard to members' subscriptions was:

1 April (beginning of year):	subscriptions owing £440
	subscriptions in advance £280
31 March (end of the year):	subscriptions owing £560
	subscriptions in advance £240

A catering account of £350 was owing on 31 March; £160 was due to the Midland Electric and £120 owing for wages.

From the above information, you are required to make up the income and expenditure account of the club and a balance sheet as at 31 March.

2 Janetta Meade and Philippa Richmond are in partnership sharing profits and losses in the proportions two-thirds and one-third. Interest at 6% is allowed on capital and Miss Richmond is entitled to a partnership salary of £400 a year before the profits are apportioned. Draw up the final accounts of the partnership from the trial balance and adjustments below:

Trial Balance 31 December

	£	£
Capital accounts 1 Jan.		
J.M.		7000
P.R.		4500
Current accounts J.M.	120	
P.R.	45	
Partners' drawings J.M.	2800	
P.R.	1700	
Partnership salaries	200	
Purchases/sales	5600	16750
Returns inwards/outwards	40	70
Wages & NIC	2250	
Carriage outward	120	
Provision for bad debts		105
Trade debtors/creditors	2400	1250
Machinery (less depreciation)	7650	
Depreciation for year	850	
Fittings and equipment at cost	460	
Lighting/heating	330	
Office salaries & NIC	1600	
		(Contd)

Stock 1 Jan.	1 500	
Warehousing expenses	230	
Advertising	400	
Rent, rates and insurance	960	
Bad debts	85	
Cash and bank balances	705	
Discounts allowed/received	325	455
	30 250	30 250

a Stock on hand 31 December £1820
b Rates paid in advance to 31 March £100
c Provision for debtors to be maintained at 5%
d Wages owing at 31 December £25
e Carry forward £200 already paid on advertising account.

31

Accounting for management

In this chapter you will learn:
- *how managers use accounts*
- *how to present the information more usefully for managers*
- *the meaning of a 'narrative' balance sheet*

Who uses accounts? Final accounts can be used by a wide range of people: the bank, to see if they should lend money; partners to split the profit between them; HM Revenue and Customs, to base their tax computation on.

But management also need the information that is in the accounts, to see how well the business is doing. Typically they need to consider the results on a monthly rather than an annual basis. Accountants and managers do not always get on well. Managers will often dismiss accountants as 'bean-counters', because accountants think in figures, whereas accountants can't understand why managers can't read financial accounts. So it is important for the smooth operation of the business that accountants provide information for managers in a way which is easier for them to understand. Look at this trading account, which by now you should see as simple and straightforward.

Trading a/c month to 30 June

	£			£
Stock 1 June	1000	Sales for month	£6500	
Purchases for month £2200		Less returns	100	6400
Less returns 50	2150			
	3150			
Stock 30 June	850			
	2300			
Production wages	1800			
Warehousing expenses	120			
Cost of sales	4220			
Gross profit c/d	2180			
	6400			6400

By way of comparison, the smaller statement shown next is even more informative with less detail to absorb, and would be easier to understand by the non-accountant.

Trading a/c month to 30 June

Net sales for the month of June		£6400
Cost of sales		
Materials consumed	£2300	
Production wages	1800	
Warehousing expenses	120	4220
	Gross profit	£2180

We could also group the various losses and expense items of the profit and loss account in a manner which would mean something more to the casual observer, thus:

	£	£
Gross trading profit		2180
Distribution expenses		
Carriage (outwards)	25	
Van repairs	36	
Petrol and oil	23	
Depreciation of vans	65	149
Selling expenses		
Advertising	52	
Postages (proportion)	14	66
Office and administration		
Salaries	490	
Telephone	44	
Insurance	18	
Printing/stationery	33	
Postages (proportion)	28	613
Financial		
Discounts	30	
Depreciation office equipment	20	50
Total expenses		878
Net profit before tax		£1302

For comparative purposes, the figures for the previous year could be shown in the marginal space on the right. Sometimes, too, each item of expense is calculated as a percentage of net sales.

To present final accounts in narrative form, the trial balance can be made up on these lines, *after all adjustments* have been dealt with and the cost of sales figures ascertained, as shown in the following example.

Trial Balance 30 June

	£	£
Fixed assets		
Premises at cost	60 000	
Van (book balance)	3 000	
Current assets		
Stock 30 June	12 300	
Trade debtors	13 000	
Rates in advance	800	
Cash/bank	5 200	
Loan capital		
10% mortgage on premises		20 000
Current liabilities		
Trade creditors		6 000
Expense creditors		400
Interest due on mortgage		2 000
Claims of proprietorship		40 000
Capital owned		
Revenue income		
Net sales		127 600
Revenue expenditure		
Cost of sales	65 000	
Salaries & NIC	22 000	
Office expenses	6 300	
Advertising	2 200	
Van expenses	500	
Rates and insurance	2 700	
Mortgage interest	2 000	
Depreciation of van	1 000	
	196 000	196 000

Note that the inclusion of the cost of sales figure on the trial balance means that stocks and purchases have already been adjusted to arrive at the total debit of £65 000. In consequence, the stock figure of £12 300 on the trial balance is that of the closing stock of 30 June. Note also that the van account shows the up-to-date balance, after the adjustment for depreciation, and that the mortgage interest, being due and not yet paid, is shown both as a debit (to be charged to revenue) and also as a credit (under current liabilities).

Revenue Account for the year to June

			£	Previous year
Net sales for the month			127 600	
		£		
Costs of sales		65 000		
Selling/distribution	£			
Advertising	2 200			
Van expenses	500			
Depreciation of van	1 000	3 700		
Office/administration				
Salaries & NIC	22 000			
Office and general	6 300			
Rates and insurance	2 700	31 000		
Financial expense				
Mortgage		2 000	101 700	
Net trading profit for month			25 900	

Insight

The narrative style of presentation, for both the revenue account and the balance sheet, has been adopted generally throughout the commercial world. In particular, this format, with its many explanatory footnotes for greater ease in interpretation, lends itself to the final accounts of limited companies. It is shown in further detail in Chapter 36.

The balance sheet in this illustration is shown overleaf, but since it refers in this instance to the accounts of a sole trader, note that there is very little detail, on the capital side, when compared with the balance sheet of a limited company.

Note that in this illustration for simplicity the proprietor's capital account has been made up to date at 30 June, presumably after deduction of any drawings for the year, but before the addition of the net trading profit.

Balance Sheet as at 30 June

Assets employed			Previous year
Fixed assets	£	£	
Premises at cost	60 000		
Van, less depreciation	3 000	63 000	
Current assets	£		
Stock 30 June	12 300		
Trade debtors	13 000		
Payment in advance	800		
Cash/bank	5 200	31 300	
Current liabilities			
Trade creditors	6 000		
Expense creditors	400		
Mortgage interest	2 000	(8 400)	
Working capital		22 900	
		85 900	
Financed by			
Proprietary capital	£	£	
Balance 30 June	40 000		
Net trading profit	25 900	65 900	
Fixed liability			
10% Mortgage loan			
(secured on premises)		20 000	
		85 900	

32

Costs of production and manufacturing accounts

In this chapter you will learn:
- *how managers use financial information for costing and pricing*
- *the difference between prime cost and overhead cost*
- *how to prepare a manufacturing account*

In the last chapter we saw how to rearrange the financial accounts into a more narrative form, grouping expenses and items on the balance sheet so that they were more easily understood by managers. In this chapter we are going to look at how the information in cost of sales can be expanded and grouped in a way that makes it more useful for managers who have to make pricing decisions.

Take the example of a widget making factory. To make one widget you need £1 of materials. A worker can make three widgets an hour and is paid £6 an hour. You might think that the cost of a widget is therefore £3 (£1 materials and £2 labour) so that if they are sold for more than that there will be a profit.

However, this does not take into account depreciation on the machinery, heating and lighting the factory, etc. Assume these fixed costs are £100 000 a year in total, and the factory produces 100 000 widgets a year. So the cost is actually £4 per widget, and they must be sold for more than that (on average) to make a profit.

Does that mean that you should turn down a large order at £3.50 a widget? Not necessarily. The more widgets you make, the less each one has to bear of the fixed cost of £100 000. In any case, if the existing 100 000 widgets are being sold at £4 each you have covered all your fixed costs, so you would still be making a profit of 50p on each **extra** widget sold at £3.50.

The complexities involved in setting prices like this mean that managers need to have a clear picture of the costs which vary with production – the **prime costs** – and those that don't – the **overhead costs**.

Prime cost and oncost

A manufacturer must buy raw material and allow for all manner of labour and factory costs long before selling the finished product, sometimes weeks, perhaps months afterwards. Meanwhile, that same manufacturer probably has to borrow money and pay interest to maintain production.

Prime cost is the term used for the direct expenses of production – the materials, labour and the actual expenses that can be definitely allocated to certain production schedules. In addition, the manufacturer incurs many other expenses that cannot, as a rule, be allocated easily to a particular job or contract. These expenses include the rent and rates of the factory, lighting and heating, factory power, the wages of supervisors, time-keepers and mechanics, and the wear and tear (depreciation) of the factory plant. These necessary and essential expenses of maintaining the factory in a productive state have to be charged to each job or contract under the heading of **oncost** or **overhead expense**, generally on a percentage basis over the whole production.

Stocks and their turnover

Large stocks held in the warehouse do not earn their keep, yet minimum stocks must usually be maintained, based upon past experience to ensure demand can be met at short notice.

A likely aim of management is to increase the rate of stock turnover, thereby increasing trading profit. Periodic checks are made by the cost accountant (in big industry) comparing the rate of stock turnover with that of earlier periods. An illustration of a small manufacturer's cost statement is now shown, providing useful information for both the cost and financial accounting.

Note how the cost of production fits into the trading account, as shown below, taking the place of the ordinary purchases of a non-manufacturing concern.

Cost of Production Statement

	£	£
Stock of raw materials 1 Jan.	8 500	
Purchases of raw material	72 350	
Carriage on raw material	150	
	81 000	
Less stock of raw materials 31 Dec.	10 000	
Raw materials consumed	71 000	
Production wages & NIC	110 000	
Direct expenses of production	2 200	
Prime cost		183 200
Factory overheads:		
Heating and lighting	1 700	
Rent and rates	3 500	
Power	2 400	
Supervision (foremen)	8 800	
Depreciation of machinery	2 000	
Overhead expense		18 400
Cost of production		201 600

Trading Account for the year ended 31 December

	Current year		Previous year	
	£	£	£	£
Net sales for year		275 000		220 000
Cost of sales				
Stock of finished goods 1 Jan.	32 000		34 000	
Cost of production b/f	201 600		158 800	
	233 600		192 800	
Less stock of finished goods 31 Dec.	36 000		32 000	
	197 600		160 800	
Warehouse wages & NIC	4 400		4 200	
Cost of sales		202 000		165 000
Gross trading profit		73 000		55 000

Rate of stock turnover

Current year	*Previous year*
$\dfrac{202\,000}{34\,000} = 5.94$ times in the year	$\dfrac{165\,000}{33\,000} = 5$ times

(approx. every two months)

The gross profit ratios in the above illustration are:

$$\frac{73\,000}{275\,000} \times 100 = 26.55\% \text{ for the current year}$$

$$\frac{55\,000}{220\,000} \times 100 = 25.00\% \text{ for the previous year.}$$

The increased ratio may be due to a variety of causes – increased prices, reduced level of overheads through greater production, or perhaps more effective control over overhead or administrative expenditure. However, the use of this kind of analysis will help the manager to highlight that her business has changed from previous periods and may need checking to ensure all is well and in control.

Manufacturing accounts

The costs and expenses at basic production level determine, to a large degree, the ultimate selling price of the finished product. These bulk costs normally occur long before the retail stage.

The manufacturer must obtain raw materials, an efficient labour force and the machinery of production involving heavy capital finance, months ahead of the estimated demand for any marketable product.

The cost of production statement (above) is also a manufacturing account, preceding and leading into the business trading account. Generally, it would comprise a greater list of debit items, all costs and expenses of production, the total being referred to as the **cost of production**. This cost of manufacture or production is brought down to the debit of the trading account we already know, taking the place of the purchases figure to which we have become accustomed. But now that we are about to make up a manufacturing account, remember that some finished goods may have been bought from outside suppliers at this stage. These are shown as an additional debit in the ordinary way.

SEVERAL CLASSES OF STOCK

In making up another more detailed manufacturing account, and processing it through to the final net trading profit stage, take note of the three different kinds of stock: raw material, partly finished goods (sometimes called work in progress), and finished goods. The latter belongs only to the trading and selling section of the business, whereas the first two type of stocks are found only in the manufacturing or production account. The closing balances, however, of all three kinds of stock are eventually taken to join the current assets on the balance sheet.

The old-fashioned style of manufacturing account has been drafted below to emphasize the various 'account' stages and sections and to draw the student's attention to the sectional terms in use – materials consumed, prime cost, factory/works cost, and the cost of sales.

Manufacturing, Trading and Profit and Loss Account
for the year ended 31 December

	£		£	
Stock of raw materials 1 Jan.	2 000	Work in progress		
Purchases of raw materials	36 000	31 Dec.	6 000	
Carriage on raw materials	1 000	*less* 1 Jan.	5 500	500
	39 000			
Less stock of raw materials		Cost of		
31 Dec.	5 000	production c/d		75 840
Materials consumed	34 000			
Direct wages & NIC	30 000			
Direct expenses	6 740			
Prime cost	70 740			
Overhead expense				
including deprec. of plant	5 600			
Factory/works cost	76 340			76 340
Stock of finished goods 1 Jan.	30 000	Net sales		130 000
Cost of production brought				
down	75 840			
Purchases of finished goods	1 400			
Carriage of finished goods	100			
	107 340			
Less stock of finished goods				
31 Dec.	25 000			
Cost of sales	82 340			
Gross profit c/d	47 660			
	130 000			130 000
Carriage outwards	700	Gross profit b/d		47 660
Salaries & NIC	8 800	Discounts		
Office expenses	4 400	received		300
Discounts allowed	440			
Advertising	3 120			
Net trading profit	30 500			
	47 960			47 960

NOTE ON THE MARKET PRICE OF GOODS MANUFACTURED

This type of examination problem is quite straightforward, provided examples have been studied. Sometimes, though, examiners like to add a footnote about the retail value or price of the goods manufactured on the open market.

If, for instance, in addition to all the information above, the examiner had introduced the figure of £90 000 as the approximate market value of the manufactured goods (shown as the cost of production amounting to £75 840), to show an estimated 'profit on manufacture' of £14 160, the simple adjustments to the manufacturing account already illustrated would be as follows.

In place of the cost of production figure of £75 840 on the credit of manufacturing account insert the current market value of £90 000. Carry down and debit this £90 000 to the next section of this account (the trading account) in place of the figure £74 840 for the cost of production.

The estimated profit on manufacture £14 160 will be debited (as the balance of the manufacturing account) and *brought down to the credit of the third section of this combined account, the profit and loss account.*

This is simply a management accounting concept to measure performance and compare production costs with the prices of finished goods on the open market. There is no change in the ultimate and actual net profit.

TESTING YOURSELF

32.1 Draw up a comparative trading account from the information given below. Ascertain the cost of sales and calculate the gross profit and rate of stock turnover for both periods.

	Current year	Previous year
	£	£
Raw material consumed	60 000	53 000
Manufacturing wages	77 000	69 000
Warehousing wages	6 000	5 300
Direct expenses	5 000	4 900
Overheads	10 000	8 800
Net sales	240 000	200 000
Stocks of finished goods		
1 January	26 000	25 000
31 December	24 000	26 000

32.2 Prepare a manufacturing account from the following figures, to show clearly:

a *the raw materials consumed* **b** *the prime cost*
c *factory overheads* **d** *cost of production.*

	Raw material	Work in progress
	£	£
Stocks { 1 July (one year ago)	2 500	450
30 June (at year end)	3 000	520

	£		£
Manufacturing wages	14 400	Fuel and power	1 500
Indirect wages	4 300	Direct expenses	430
Works manager's		Royalties on production	350
salary	3 000	Deprec. of machinery	200
Raw material		General maintenance	150
purchases	18 700	Insurance of factory	100
Factory rent		Carriage on raw	
and rates	2 200	material	120

32.3 You are required to make up a combined manufacturing, trading and profit and loss account from the following information, and give details of:

a *materials consumed* **b** *cost of production*
c *profit on manufacture* **d** *cost of sales*
e *net profit for the year.*

	Raw material £	Finished goods £	Work in progress £
Stocks { Jan. 1	1000	2375	1900
{ Dec. 31	625	2760	1750

	£		£
Salaries	3 350	Carriage outwards	410
Office expenses	1 190	Raw materials purchases	7 500
Factory expenses	3 188	Carriage on material	125
Discounts received	350	Sales less returns	49 000
Advertising	1 800	Warehousing expenses	865
Discounts allowed	300	Purchases/	
Factory wages	12 110	finished goods	1 175

Factory plant and machinery is to be depreciated by £2750. The market value of the actual cost of production at the balance sheet date is estimated at £30 450. This is to be taken into account in drafting the manufacturing account.

33

Introduction to limited companies

In this chapter you will learn:
* *what a limited liability company is*
* *how it raises capital via issuing shares and taking loans*

Limited liability

The modern form of **limited company** came into common use during the last half of the nineteenth century. The main form of trading vehicle before that had been the sole proprietorship or the partnership, but both of these are unwieldy when the people putting up the capital are different from those actually managing the company. Ostensibly this is the main reason for a limited company – so that someone putting up capital for a venture without the intention of being involved in managing it will only have that limited amount of capital at risk, rather than having unlimited liability against all personal wealth. The limited company is a **legal entity** – a legal 'person' – in its own right, and can be sued for debts owed, breaches of contract etc. But it is a different person from the shareholder or the director, and neither can (in principle) be required to pay money which is owed by the company.

This principle holds true for the large companies quoted on the stockmarket, where the shareholders are entirely different from the company and from the directors. However, it has to be said that of the 'live' companies currently listed at Companies House, only a small percentage will have any real separation between owners and managers. The vast majority will have shareholders who are also the directors of the company. In spite of this they are still able to maintain the legal position that they are not personally responsible for the debts of the company. The company here is a vehicle for providing limited liability from the creditors of the trade that they themselves carry on through the company, and a means to manage the tax liability on undistributed profits.

Regardless of the economic reality, the law and accounting procedures governing a limited company rely on the principle that ownership and management are two entirely different functions.

Owners and managers

The owners of a limited liability company are the **shareholders**. They each own a number of shares in the company, and their entitlement to share in the profits of the company depends on the proportion of the total issued shares that they own.

The shares will normally carry voting rights, again in proportion to the number of shares owned. So if a company has issued 100 ordinary shares, 80 of them held by one person and 20 other people each holding one share, the shareholder with 80 shares can outvote all the other 20 put together.

The shareholders, meeting in an annual meeting, will elect directors to manage the company on their behalf. These are the managers of the company, although some may only have a part-time role centred on attending board meetings (known as **non-executive directors**).

It should be apparent that the role of the financial statements in this respect is crucial, since they are the main way in which the shareholders, who are not involved in the day-to-day operation of the company, can see how the managers are carrying out their job. Because of this, there are detailed statutory rules on how the accounts of a limited company should be set out. These are contained in the Companies Acts, passed from time to time by parliament, and sometimes in other regulations and rules.

Additionally, limited companies may have to have an annual **audit,** where their financial statements are checked by an independent (to the business) firm of accountants, who give an opinion as to whether they give a true and fair view of the business's financial position and recent activity on behalf of the shareholders. Smaller companies are not required to have an audit although may need to have some sort of checking of their accounts depending on their status as a business or if the law requires it of them.

Capital of a limited company

The constitution and rules of a company are contained in two documents called the **Memorandum** and **Articles**. These set out the types of share capital the company is entitled to issue, and the rules governing the voting of shareholders and the relationship of shareholders to the company.

In the simplest type of capital structure, a company has the right to issue a given number of ordinary shares (its **authorized share capital**) which it has actually issued perhaps only partly (called its **issued share capital**) to shareholders, who have paid in full the price put on the share by the company. For example, a company might have an authorized share capital of one hundred £1 shares which it has issued equally to two shareholders, A and B, who have each paid £50 for their fifty shares. The company has an authorized, issued and fully paid share capital of £100. If the company were now to be unable to pay its debts, the shares would become worthless, but A and B would not have to pay out anything further.

The most common variation from this is that the issued share capital is less than the authorized share capital. So in the above example, A and B might only have taken one share each, paying £1 each, with no other shares being issued. They would still each own half of the company, and are each entitled to half of the profits from it, but the company would only have £2 in its bank account instead of the £100 it raised previously. The shares are still fully paid, and if the company could not pay its debts neither A nor B would be obliged to pay any more of their money in. The only difference that having the unissued shares makes is that the company does not have to alter its memorandum if it wants to issue more shares.

Partly paid capital is another matter. Instead of the last example, A and B might have each taken the fifty shares, but only paid 2p in the pound for each of them. This is called partly paid capital. 2p each for fifty shares means that they would have paid £1 in total just as in the last example, and they would each own half of the company and be entitled to half of the profits. But this time, if the company were unable to pay its debts, A and B would be liable to pay the outstanding capital of 98p a share, giving them each a liability of £49. This would, however, be the limit of their liability. It is comparatively rare to have partly paid share capital except where the price of shares is paid in instalments in a public issue.

Shares give the shareholder a right to receive a share of the profits, but only when it is agreed between the shareholders to distribute them – there may be times when no distribution is made for a variety of reasons (insufficient profit, need to reinvest profits into the firm to enable it to grow and so on). This distribution is called a **dividend** and will be declared as so much per share. The decision on payment of a dividend is taken at the Annual General Meeting (AGM), although in practice it is usually the directors who decide how much is to be paid out by proposing a dividend payout for the year that the shareholders will then ratify as part of the AGM. Shares can normally be bought and sold freely between those owners willing to sell them and other people who want to buy them (either as other existing shareholders or otherwise), but they are only worth whatever someone is prepared to pay for them. Unless they are specifically issued as redeemable shares (which is not normal) they need never be repaid by the company.

Loan capital

In addition to raising money by issuing shares, companies may also raise money by borrowing it. Sometimes they will simply borrow from a bank, but sometimes they will issue loan certificates, known as **debentures**, sometimes called loan stock. These normally carry a fixed rate of interest, and a right to have the capital repaid at a certain time.

For example, a company might issue £10 000 8% debentures redeemable in 2020. That means that it will receive £10 000 from the people who originally buy the debentures, but every year until 2020 it must pay the holders interest totalling £800. In 2020 the £10 000 must be repaid to the debenture holders. This is regardless of whether or not profits are made by the company; on the other hand, the debenture holders have no right to any further income regardless of how well the company performs.

The accounting entries for the issue of various types of capital are considered in the next chapter, but in the meantime it is important to understand the differences between the financial statements of a limited company and a sole trader.

Characteristics of limited company accounts

There are two main areas of difference – what happens after net profit in the profit and loss account, and what is shown in capital on the balance sheet.

Having calculated net profit as for sole proprietors, this is taken down to an appropriation account, which shows how the profit is to be appropriated.

The first thing to be taken off is *corporation tax due*, to give *profit after tax*. Next any *balance brought forward* from previous years is added. The *dividends proposed on the shares* are then deducted, and finally a *balance of profit brought forward* is left. It is illegal for a company to distribute more than it has in total profit, both earned in the year and brought forward.

EXAMPLE

Limited company appropriation account	
Net profit	10 000
less corporation tax	2 400
	7 600
add retained profit b/f	4 000
	11 600
less dividend proposed	8 000
Retained profit c/f	3 600

On the balance sheet, the share capital appears separately from the brought forward profits, but all of them are still shareholders' funds. The brought forward profits are sometimes called reserves, although there are also other credit balances that can be referred to in this way.

34

Accounting for share capital

In this chapter you will learn:
- *the difference between registered and issued share capital*
- *how to account for share capital*
- *what happens when a shareholder defaults on a call for capital*

Registered and issued share capital

The nominal, authorized or registered share capital is that stated in the company's charter, the Memorandum of Association. That part of the authorized capital which is issued to the public is shown under the heading of issued capital on the balance sheet in this manner.

Authorized and issued share capital		
Authorized capital	£	£
100 000 10% preference shares of £1 each	100 000	
400 000 ordinary shares of 50p each	200 000	300 000
Issued capital		
100 000 10% preference shares of £1 each fully paid	100 000	
300 000 ordinary shares of 50p each fully paid	150 000	250 000

Note that it is only the issued share capital which forms part of the double entry system, and consequently the authorized capital is ruled off when it has not been fully issued.

There are various categories of capital. In the figures shown above, the nominal, registered or authorized capital is £300 000. The issued capital is £250 000, consisting of 100 000 10% preference shares fully called and paid up, and 300 000 ordinary shares of 50p called and paid up. The uncalled capital of 100 000 ordinary shares of 50p has not yet been issued to the public.

There are also various classes of shares, the two main classes being preference and ordinary. The former carries a fixed rate of interest which is payable, dependent upon the company's available profits, *before* the payment of any dividend recommended by the directors to ordinary shareholders.

Share issues

With some public share issues, payment for the shares taken up may be made by instalments, in the following manner.

A new company is 'floated' with an authorized share capital of 200 000 ordinary shares of £1 each. Half of the registered capital is to be offered to the general public and to be paid in three distinct stages, thus:

> *20p on application by 30 April*
> *30p on allotment on 1 June*
> *50p first and final call on 1 July.*

The build-up of the company's finances is now shown over these two months. No complications are assumed (refunds, unpaid calls, etc.). The money comes in direct to the company's bank account, and the ordinary share capital is built up this way, via application, allotment and final call accounts:

CASH BOOK Bank receipts (debit) side

	30 April	1 June	1 July	Total
	£	£	£	£
Application a/c	20 000			20 000
Allotment a/c		30 000		50 000
Call a/c			50 000	100 000

Ordinary Share Capital (credit) side

	30 April	1 June	1 July	Total
	£	£	£	£
Application a/c	20 000			20 000
Allotment a/c		30 000		50 000
Call a/c			50 000	100 000

Assuming that no trading operations have yet taken place, the company balance sheet on 1 July appears as follows.

Assets of the company		
Bank account		£100 000
Represented by		
Authorized share capital		
200 000 ordinary shares of £1 each	£200 000	
Issued share capital		
100 000 ordinary shares of £1 fully paid		£100 000

Application and allotment

This is a general guide to the double entry procedure.

		Debit	Credit
a	Completed forms for the purchase of shares and application money received from the general public.	Bank	Application account
b	Company confirms allotment to successful applicants –	Applic. a/c	Share capital a/c
c	and asks for amounts due on allotments.	Allot. a/c	Share capital
d	Some money may be returned to unsuccessful applicants with letters of regret.	Applic. a/c	Bank
e	Allotment money comes in.	Bank	Allotment account
f	If some money is unpaid (owing) on allotment.	Calls in arrear	Allotment a/c
g	The directors decide to make a call for further money.	Call account	Share capital
h	The call money comes in.	Bank	Call account
i	If some calls are unpaid (owing).	Calls in arrear	Call account

Application and allotment accounts are generally combined.

Note that after the application stage, the share capital of the company is *credited before the money is actually received* on the allotment and call accounts.

Calls in arrear

The liability of a company member is limited to the amount he (or she) has contracted to pay for the shareholding. If he has not paid all the 'calls' made on him by the company, he becomes a debtor for the money that is due. Sums owing by a defaulting shareholder are referred to as **calls in arrear**.

If, in the last illustration, one shareholder (John Brown) failed to pay the final call of 50p on his allotted 500 shares, he would become a debtor of the company for £250 (50p on 500 £1 shares).

The share capital at 1 July would show the full credit balance of £100 000, but cash would be short by £250.

In making up the company's balance sheet at 1 July, John Brown's debt would not be included in the ordinary trade debtors' total in the current assets, but would be deducted as 'calls in arrear' from the issued share capital in this manner:

Assets of the company		
Bank account		£99 750
Represented by		
Authorized share capital		
200 000 ordinary shares of £1 each	£200 000	
Issued share capital		
100 000 ordinary shares of £1 fully paid	£100 000	
Less calls in arrear	250	£99 750

The directors of the company will give John Brown notice that unless he pays his final call of £250 by a certain date, his shares will be forfeited and sold (probably to another member). No refund would be made of the amount he has already paid.

Premium on shares

A successful company, perhaps needing further capital for expansion, might put out an additional issue of shares at a **premium**, i.e. at an amount above the nominal or par value of the shares. If, for instance, the nominal value of the shares was £1 and they were issued at £1.20 per share, the premium would be 20%. An applicant for 100 shares would pay £120 for them.

The Companies Acts require amounts received as premiums on shares to be taken to a share premium account as a capital reserve. This kind of reserve can only be used for special purposes (such as off-setting capital losses) and is not available for transfer to the credit of the revenue account and to be used for payment as a dividend, as in the case of a general reserve.

Normally the premium is paid with the allotment money, the book entries, in the first instance, being the full debit to the allotment account, with the separate amounts being credited to share capital and the share premium account. Then, when the allotment money (and the premium) is received, the bank is debited and the allotment account credited with the full amount of the allotment money plus the premium.

The premium, as a credit balance in the books, is shown on the balance sheet under the separate heading of *capital reserve*, between the issued capital and the revenue reserves, thus:

Authorized and issued share capital		
200 000 ordinary shares of £1 fully paid		£200 000
Capital reserve		
Share premium account		10 000
Revenue reserves		
General reserve account	£25 000	
Profit and loss undistributed balance	8 000	33 000

Debentures

A debenture issue is similar to a share issue, often by instalments, and it is referred to as **loan capital**. Generally, the total figure for debentures is deducted from the total of the *net assets* to give the net worth or equity of the company (see the end of the next chapter).

TESTING YOURSELF

34.1 The Colby Mail Order Company Limited was registered on 1 January with an authorized share capital of £200 000 in ordinary shares of £1. On 1 February, one-hundred thousand shares were offered for public subscription by instalments as follows:

On application	10p per share
On allotment	20p per share
On 25 March	30p on first call
On 20 June	40p second and final call

The issue was fully subscribed. All monies were received on the due dates with the exception of the second and final call of one member, a Mr Frank Evans, who had been allotted 1000 shares.

Journalize the accounting entries and make up the balance sheet of the company, in so far as these particulars are concerned, as at 20 June.

34.2 Jaybee Enterprises Ltd issued 200 000 Ordinary Shares of £1 at a premium of 10%, and also £60 000 12% debentures at par.

Applications were received for 212 000 shares. The money surplus to the company's requirements was sent back with 'letters of regret' to those applicants who had asked only for small batches of shares.

The shares were received in three instalments: 20p on application, 40p on allotment (including the premium), and 50p two months after allotment. All monies were received on the share issue except in the case of one shareholder of 400 shares who could not meet his final call. The Board decided to allow this member another month to pay the amount outstanding, otherwise his shares were to be forfeited.

The debenture issue was subscribed in full and all monies banked.

Write up the cash book of the company and all relevant ledger accounts, and show the balance sheet, incorporating these details, made up on 30 September, the day after the date set for the final call.

35

The final accounts of
a limited company

In this chapter you will learn:
- **the statutory requirements for company accounts**
- **the importance of disclosures on the balance sheet notes**

This chapter, on company accounts, is still introductory and simply intended to cover the requirements of the first-level Accounting examination syllabuses.

The 'published section' of the profit and loss account

The Companies Act 1985 provides a rigid hierarchy of headings which must be used in reporting the results of a limited company. The main headings for a profit and loss account, for example, begin as follows:

- ▶ *Turnover*
- ▶ *Cost of Sales*
- ▶ *Gross profit or loss*
- ▶ *Distribution costs*
- ▶ *Administrative expenses.*

The Companies Acts also require full public disclosure of certain items which are normally included as notes. Full details must be given of directors' fees and salaries, loan and debenture interest, and the depreciation of fixed assets.

Information must also be given of the following 'appropriations' of profit:

- ▶ *amounts paid or reserved for taxation*
- ▶ *dividends paid or recommended*
- ▶ *amounts transferred to (or withdrawn from) reserves.*

The purpose of these disclosures is for the benefit of the shareholders, creditors, debenture holders and HM Revenue and Customs.

Balance sheet disclosures

The format for the balance sheet is equally rigid, for example, with Fixed Assets normally forming the opening item, split into intangible, tangible and investments, and each of these further divided into four or more categories.

The detailed requirements of the Companies Acts need not be memorized at this level of study and the next few paragraphs are included by way of interest.

The authorized share capital, where it is a different amount to the issued share capital, is simply presented for information and ruled off as it does not form part of the double entry.

Capital and revenue reserves are to be shown under separate headings. The former cannot be paid away in dividend, whereas the latter can be written back to the credit of profit and loss appropriation account if recommended by directors and approved by members.

Long-term liabilities are to be shown under their own headings. Current liabilities will include any bank overdraft, loan and debenture interest owing, expense accruals, and *dividends proposed but not yet paid*.

Notes detail any arrears of dividend of certain types of shares (cumulative preference shares); also details of any contingent liabilities, and capital expenditure authorized by the directors.

A company's fixed assets are shown at *cost less aggregate depreciation written off*. The fixed assets are grouped in order of permanency (land, buildings, plant and machinery, fittings and fixtures, motor vehicles) and totalled separately from the current assets.

Current assets are generally shown in this order: stock on hand at the date of the balance sheet, trade debtors, pre-payments such as rates in advance, and the bank and cash balances.

Notes are to be made of acquisitions or disposals of fixed assets during the year, and a statement made of the method of arriving at the valuation of the stock of finished goods and the work in progress.

The vertical presentation of the balance sheet discloses the working capital, net assets and the equity of the company at a glance.

The auditor's report, if there is one, is part of a company's annual report and accounts, confirming that they depict a 'true and fair view' of the company's financial affairs and comply with the requirements of the Companies Acts. A qualified report is issued when the auditors are not satisfied about some issues, and is generally a warning for shareholders to ask questions.

Auditors do not prepare the final accounts of the company and are not employed by the company. They belong to a completely separate and distinct firm of professional accountants.

A full example of published company accounts has not been included here, as it would have to be over-simplified to reproduce it.

Instead, students are urged to look at the financial statements of quoted companies to see how they are laid out.

Insight

The easiest way to see large plc company accounts is to view them online. Almost all of the larger businesses in the UK have websites with their final accounts on them to view or download for analysis. Simply use a search engine like Google to find the Investor Relations area of the website for the company you are interested in. If you prefer to have them printed out for you, the *Financial Times* carries details on its share price page of a free service for obtaining copies of accounts.

TESTING YOURSELF

35.1 Make up the profit and loss account, including a separate 'published section', of the Kaypee Trading Co. Ltd from the following information:

	£
Turnover (net sales) for the year ended 31 December	380 000
Gross trading profit for the year	168 900
Undistributed profit from previous year	4 500
Rent and rates	8 600
Bad debts written off	460
Insurance paid	380
Commissions received	7 400
Directors' fees	10 000
Salaries of directors	16 000
Office salaries and NIC	37 500
Depreciation of vehicles	3 250
Half year's preference dividend paid June 30	5 000
Advertising and publicity	3 520
Audit fee payable	2 000

Adjustments are to be made for the following:

Payments in advance: rates £1400; insurance £80
Commissions earned, not yet received £720
Provision for Corporation Tax £40 000
Increase of bad debts provision by £150 to £1000.

The directors recommend the payment of the second half of the preference dividend, and the transfer of £8000 to general reserve. They further recommend a dividend of 15% to ordinary shareholders amounting to £30 000, and ask for the approval of members for the writing off of one-third of the formation expenses debit, standing in the books at £4800.

35.2 The MM Company Limited was registered with an authorized capital of £50 000 in 100 000 ordinary shares of 50p each. All the capital had been issued and was fully paid.

You are required to draw up the balance sheet of the company from the information below.

	£
Plant and machinery (cost £30 000; depreciation to date £12 500)	17 500
Van (cost £3400; depreciation to date £1850)	1 550
Freehold premises at cost	34 000
Credit balance on profit and loss of undistributed profit	5 660
General reserve account	4 000
Trade debtors and creditors £6400:£4230	
Wages owing £440 Rates in advance £320	
Stock on hand at 30 June	16 500
Bank overdraft at 30 June	300
Provision for bad debts	640
Provision for Corporation Tax	10 000
Investments held by the Company (9% Treasury Stock, present market value £2800)	3 000
Ordinary dividend recommended by directors at 8%	4 000

36

Accounting standards

In this chapter you will learn:
- *what accounting standards are*
- *the role of the Accounting Standards Board*
- *the international dimension to accounting standards*

Most of this book sets out one way, and one way only, of entering a particular transaction. However, if you move to higher level examinations you will find that for many accounting transactions there are potentially several different ways in which they could be recorded.

For example, in all the examples given so far, you have been provided with a figure for depreciation. But how should the rate of depreciation be set? Is it acceptable to change from one method of depreciation to another? Should buildings be depreciated, on the grounds that they eventually become uneconomic to repair or adapt and have to be torn down and rebuilt, when in practice the rise in value of the land they stand on has tended to outstrip the loss in value of the building itself?

It is important that there is consistency between organizations, especially quoted companies, in the way that such transactions are recorded. If there is not, there is no realistic way that the financial statements can be used to compare one with another unless a great deal of background information is given and analysed.

Accounting standards

While legislation may provide some uniformity, the accountancy profession has also responded by setting up the **Accounting Standards Board** (ASB) which issues binding guidance on how certain transactions should be accounted for and reported. From the 1970s onwards the ASB produced 25 **Statements of Standard Accounting Practice** (SSAPs) on subjects as diverse as disclosure of accounting policies, accounting for depreciation, and accounting for goodwill.

In 1990 the composition of the ASB was changed, making it more independent of the accountancy institutes which set it up. It now issues **Financial Reporting Standards** (FRSs) and Urgent Issue Task Force (UITF) Abstracts, and the intention is that these will eventually replace all the existing SSAPs. Until then, some SSAPs remain in force.

It is often the deliberate policy of the FRSs to issue statements of principle rather than specific accountancy treatments. This seems to have come about because companies and their auditors were attending more to the letter of the SSAPs and not to the principle that they were trying to enforce. Thus for example FRS 5 lays down the general principle that *the substance and not the form of a transaction should be reported in the balance sheet*, and then gives guidance on how that should be implemented, whereas SSAP 21 (although it still remains in force) represents an earlier attempt to get certain equipment lease contracts reclassified as purchases. Whereas SSAP 21 could be avoided by rewriting the terms of the lease, it is less easy to avoid the general principles involved in FRS 5.

International standards

Whilst it is important to compare accounts of companies within the same country, it is also important to compare them on an international basis, and the importance of this increases as the internationalization of economics increases. In 1973 the **International Accounting Standards Committee** was formed to harmonize standard-setting internationally. This became the International Accounting Standards Board in 2001. Their remit is to create a single set of International Financial Reporting Standards (IFRS) that can be applied worldwide. From 1 January 2005 European listed companies are required to use IFRS when compiling their financial statements so many larger business now comply with these standards.

Insight

IFRS are in effect global standards for accounting as more than 100 countries now use them for their own national accounting standards for larger businesses at least (in fact the only major country that does not use IFRS now is the USA – but many of the principles used by US regulations of accounting often follow very similar principles to IFRS, and are getting closer to each other all the time). It is highly likely that we will eventually have one set of internationally agreed accounting standards used right across the world.

Smaller companies

An increasingly relevant issue as accounting standards proliferate is their applicability to smaller companies. Small companies and their accountants complained that the standard setting process was geared to rooting out sophisticated massaging of the figures by large quoted companies, yet the standards are applied with equal rigour to small companies that simply do not need them.

The ASB approached this by drafting a single standard for smaller companies that includes the key elements of the other standards but excludes some of the more obscure and irrelevant sections. It is known as the FRSSE (Financial Reporting Standard for Smaller Entities) and is the only standard which applies to a majority of owner-managed companies.

37

Non-financial reporting

In this chapter you will learn:
* *the limitations of financial accounts*
* *about attempts to introduce non-financial reporting*

It is appropriate near the end of the book to point out the limits of traditional accounting. The opening page pointed out that accounting was the *arithmetic of commerce* and as a result accountancy finds it hard to deal with anything that cannot be reduced to financial figures.

Nevertheless, it is clear that there are other concerns which someone reviewing the activities of a company or other organization might have. Many investors are now concerned to ensure that they only invest in companies that do not pollute the environment. Trade unions will want to ensure that any investments made by their pension or hardship funds only go to companies with good labour relations. There is a wide and growing range of factors which many experts feel should be capable of reflection in the accounting process. These concepts are given the general term **social accounting**, although there is little agreement on how broad the spectrum covered by the term really is.

It is still the case, however, that this is a concern primarily at the academic and research level rather than one which is always being put into practice. Many businesses consider that labour relations are still the remit of the Human Resources Department not the Finance Department. Companies are perhaps more likely to undertake a review of energy efficiency because they want to save money than because they want to save the planet from global warming. However, there is a growing belief that in the foreseeable future some mechanism will have to be found that imposes a genuine requirement on businesses to account for their actions in more than financial terms. In any case, the non-financial aspects of a business can have important financial consequences long-term that are not recognized in the normal process of accounting.

Two of the main areas where social accounting is relevant are labour relations and the environment. These are considered below.

Labour relations

Businesses will frequently say 'we are our people' – the business is only as good as the people it employs. In some businesses this is quite literally true: a company writing software for computer games may be totally dependent on the skills of its programmers. If they were all to leave, the company would be left with nothing but a few second-hand computers, and possibly some royalties from previous programs.

It may be thought strange, then, that traditional accounting actually discriminates against good employers in the short term. Consider two companies writing games software – Setendo Ltd and Ninga Ltd, each employing ten programmers on the same market rate of pay. In the first year they each produce a program, and the programs sell equally well. However, Setendo lays off half of its work force as the project nears completion after nine months, because it only needs five of them to finish it off. Ninga retains all its staff, even though there is not enough work for them in the last three months. At the beginning of the next year each company starts to write another program, and Setendo rehires five more programmers to bring the total back up to ten. This pattern is repeated for several years. What is the accounting outcome?

Initially Setendo Ltd will probably report better results than Ninga Ltd – its wage bill will be lower because it lays off staff when they are not working at full stretch. However, in the long run it may find it more and more difficult to hire good staff. Word will get round in the industry that you are far better off working for Ninga, because you don't have the threat of redundancy hanging over your head every year. Gradually the best programmers will likely end up at Ninga and the second best at Setendo, Ninga will start producing better games than Setendo, and will sell more copies and make more profits.

Of course this idealized scenario does not always work out neatly. There may be such a shortage of jobs and so many good programmers that Setendo will always be able to recruit good staff, and will always make higher profits on a traditional accounting basis. However, this does not mean that there are no adverse consequences. The stress caused by the continual layoffs may mean that some of the programmers become ill, requiring medical help from the state. Being unemployed for three months a year means that programmers will be claiming social security benefits. It may cause problems in their family lives. In general it can perhaps therefore be said that Ninga Ltd's policies have a greater social benefit to the community at large than Setendo's do. This cannot be recognized properly in financial statements. Of course, conversely some programmers may prefer to work for only part of the year but often other commitments make this lifestyle difficult to maintain long term, for example, it is harder to get a mortgage without a regular income stream that is guaranteed.

Occasionally there are attempts to introduce some form of labour relations recognition into financial reporting, for example by requiring businesses to state the percentage of their workforce which is disabled. Partly this is ineffective because companies come up with stock phrases and words that limit the effectiveness of such disclosures, but mainly it is ineffective because the overwhelming majority of investors and other users of the accounts pay no attention to it.

Environmental reporting

There is perhaps a greater interest in environmental measures because it can have an impact on investment. Although the number of investors and investment funds which will not invest in environmentally unsound companies is still small, it is growing, and there are now research bodies who analyse the results of companies to produce information for environmentally-aware investors to act on. They also produce information about other ethical issues such as involvement in the arms, tobacco or alcohol trades, any or all of which may be important to ethical investors. The perception of a company as environmentally friendly may also help sell its products, as consumers are more aware of the ecological impact of their purchases.

Again the implication may be both long-term financial to the company or a cost to the community at large. A company which uses a non-renewable source of wood for its products may make higher profits in the short term because it is the cheapest source of supply. In the long term it may find that as the supply dries up the price goes up, but even if the supply is so substantial that it outlasts the company's need for it, the environment as a whole will have suffered from the loss of the trees. The former will eventually be recognized in the accounts, but will not impact in the early years; the latter would not be recognized at all.

The main way in which environmental concerns are being brought into financial accounting is by putting a financial cost on them. By raising the taxes on car and lorry fuel, governments can encourage businesses to limit the amount of pollution they cause by travel. Providing subsidies for rail freight can switch goods from road to rail, even if road would otherwise be cheaper for the company. Imposing a tax on landfill encourages companies to recycle more of their waste, and to put pressure on their suppliers to reduce packaging etc.

Other aspects of social accounting

Environmental concerns and **labour relations** are two of the biggest elements of social accounting, but they are not the only ones. Good relations with a local community is another area often commented on – does the company involve itself with local schools and voluntary organizations? Is it aware of its responsibility to the local community not to close down a factory and create high employment? Again this can have a financial effect on the company long-term in the retention of labour and goodwill among local customers, and it can also have a financial impact on the community as a whole.

There is also a wider aspect to social accounting on a national or international level. Growth in the economy is generally reckoned to be a 'good thing', and is certainly recorded as such in the national accounts; it produces a higher rate of **Gross National Product**, or GNP. However, it can be argued that continued growth cannot be sustained without excessive damage to the environment and to health by pollution, and that we have insufficiently sophisticated systems for measuring this.

38

Computerized accounting

In this chapter you will learn:
- *how computers are used in accounting*
- *what to consider when introducing computerized accounts*

Whilst this book has talked about entries being written up in ledgers, in practice most businesses these days will keep their accounting records on a computer. However, this only increases the importance of understanding the underlying processes, because they are hidden from view inside the computer's software.

From their inception computers have been used for accounting purposes. When the only computers available were mainframes that required large air-conditioned rooms, accounting was probably the first commercial (as opposed to military) use to which they were put.

Spreadsheets

The radical change was the introduction of the desktop PC in the early 1980s. Bundled with the first PCs was a word-processing package and a spreadsheet program. A spreadsheet is a grid of cells, rather like the squares on a map. They are numbered vertically and assigned letters horizontally, so that each cell can be referenced – e.g. C5.

Numbers are then entered into some of the cells, and formulas into others. The formula might say, for example 'add together the cells from C5 to C10'. The program will then automatically calculate the answer to the formula.

The great advantage of spreadsheets is that when the entries are changed, the spreadsheet will automatically update itself so all subsequent computations stay correct. If the same calculation was to be performed manually it would take hours to make the changes.

Computerized accounting systems

It did not take long before the advantages of computerized accounting became obvious. Various systems based on special stationery with carbon paper backing had been devised to allow the double entry to be written out only once but entered on two or more accounts – the most popular was called the Kalamazoo™ system. Now computers could do the same, only far more efficiently, and could also perform all the calculations.

Typically a computerized accounting system will be modular. At the most basic level, there is little more than an extended cash book. A simple system would include general, sales and purchases ledgers. The invoicing and ordering can then be integrated by buying further modules, and payroll can be added. Ultimately the whole of the accounting function can be computerized, and the latest trend is towards integrating information from accounting into 'know your customer' systems for sales and managerial staff.

However, although the transactions are processed much faster, it is important to ensure that they are entered correctly. There is far less scope for finding errors as the calculations are carried out than there is in a manual system, where the bookkeeper might well notice that a figure seems unusually large. Computerized accounting systems therefore have carefully designed entry screens, which will reject the wrong sort of entry (trying to put text where the amount should be, for example).

Issues to consider when introducing a computerized accounting system

1 *Do you want to buy a package or have a system designed for you? It is rare now to have a completely new system designed as the costs are prohibitive, but some systems can be tailored more to your way of operating. The advantage of this is that it is easier to get used to, the disadvantage is cost (both now and of maintenance) and that you may be reproducing bad practices carried over from your manual system.*

2 *Are you going to run it alongside your manual system for a while or are you going to switch over to the computerized system immediately? It is best to run the two alongside each other until the glitches are ironed out, but this can be complex (you only have one set of source documents, for example) and costly in terms of staff.*

3 *Are you going to try and implement it yourself, or get advisers to do it for you? The latter is cheaper and means that you have a good grasp of how the system operates, the former is less stressful on staff and management and should be implemented faster.*

Answers to assignments and revision exercises

1.1 Assets = £70 000

1.2 Capital = £40 000

2.1 **a** Credit note for a sales return – a customer returning goods to the business and reducing the amount that they owe.
b Credit note for a purchase return – goods returned to the supplier, reducing the amount owed.
c Accrued expense – operating expenses owing at the end of the period, e.g. telephone bill not yet paid.
d Invoice – a document setting out details of the goods or services provided, the name and address of the purchaser and the supplier, the amount due and normally a date or time limit for payment.
e A dishonoured cheque – a payment that does not clear.

3.1 If you add up the totals on the cash account, the debits should be £455.60 more than the credits.

3.2 Answers could include knowing what is owed by the business, knowing what is owed to the business, being able to calculate profit, reporting to the tax authorities, comparing one year with another, checking bank statements are accurate, etc.

3.3 Cash sales – payment received immediately for goods sold.

Credit sales – goods sold, customer to be invoiced and pay within 30 days.

3.4 Cash purchase – paid immediately for goods purchased.

Credit purchase – receive goods and invoiced to pay supplier within 30 days.

3.5   Trade receivables (debtors) – customers who owe the business money.

Trade payables (creditors) – suppliers to whom the business owes money.

4.1   The right-hand (credit) side shows money going OUT; the left-hand (debit) side shows money coming IN. If you are confused by the fact that your bank statement is 'in credit' when you have money in it, remember that this is a copy of your bank's accounting records. A 'credit' balance means that you are a creditor of theirs, in other words they owe you money.

4.2   The balance brought down should be £1171.50.

4.3   The balance brought down should be £3161.40.

5.1   The answers to parts a–e can be found in the text. f The balance brought down should be £923.

5.2   The purchase on 30 January does not enter the cash book as it has not yet been paid for. The cash book should balance to £99.97.

5.3   **a**   A decrease in a liability: DR liability CR cash/bank.
   b   An increase in revenue: DR cash/bank CR Sales.
   c   An increase in an asset: DR asset CR cash/bank or creditors/loan.

6.1   The capital balance c/f is £720.

6.2   The cash book balance c/f is £48.20. The trial balance totals £501.50 on each side.

6.3 Salaries are a business expense, deducted from profit. Drawings are not: they represent what the proprietor does with the profit after making it.

6.4 Checking the arithmetical accuracy of the postings, and rendering the work of the accountant or bookkeeper easier in the making up of financial summaries, often with a view to the preparation of revenue trading accounts, together with a balance sheet.

6.5 To ascertain whether the trial balance has a transposition error, try dividing the difference by nine. If it divides exactly, you may have made a transposition error, for example entering 1985 rather than 1895.

Revision exercise 1 – the cash book balance c/f is £270; the trial balance should total £725.84 on each side.

7.1 The cost of sales is £4300; the gross profit £3000 (closing stock is 1400 items at £1.25 each, £1750).

7.2 The gross profit is £2350.

7.3 Cost of sales is £4400. This is £400 less than the purchases because the stock on hand at the end of the month is worth £400 more than the opening stock.

8.1 The gross profit on the trading account is £1570, the net profit is £1180.

8.2 The gross profit is £2570, the net profit is £1920.

8.3   **a**   Carriage inwards are delivery charges on purchases and a cost of sales expense.
    b   Carriage outwards is the cost of distribution and a selling expense in the profit and loss account.
    c   Drawings are amounts taken out of the business for personal use.

9.1   The trading profit is £4850.

9.2   The trial balance should total £6650, gross profit is £1073 and net profit £510. Opening capital is £2000.

9.3   The balance sheet is a financial statement, grouping and listing the business property and the capital and liabilities on that certain date.

9.4   Retained profit is that balance of profit remaining after all appropriations (or withdrawals in the case of a sole trader) have been deducted.

10.1   The cash book should show £104 c/f. The trial balance totals £2463. The gross profit is £477 and the net profit £314. Capital of £1714 is matched by the van at £650, fittings at £585, stock at £375 and cash at £104.

11.1   Cash on hand at the end of the month is £353, cash at bank is £5134. Note that receiving the invoice from ABC Services Ltd does not have any effect on the bank or cash position!

11.2   Cash on hand £100 (because the final line of the question tells you so), cash at bank £5165.

11.3   **a**   Cash book – a ledger to deal with a bank account.
   b   Overdraft – a credit balance where the bank is owed money.
   c   Contra entry – entries that cancel themselves out, transfers between office cash and bank.

11.4   The important thing is to avoid the complication of an overdrawn cash balance which, strictly, should not be possible. When in doubt, pay by cheque.

11.5   Entries on the credit side of the cash book originate from the official receipts from suppliers, cash memos from sundry small purchases, and the counterfoil stubs of the firm's cheque books.

12.1   £728.24 – £224.75 (not yet cleared) + 82.64 + £110.80 (not yet presented) = £696.93

12.2   Debit balance of £22.84.

12.3   This is due to the time lapse between the receipt of cheques from the firm's customers and their clearing from the bank accounts of the debtors. Again, there is a similar delay between the handing over or posting of cheques to suppliers or creditors and their collection, via their bankers, from the account of the paying firm. Bank charges, direct debits and standing orders also need to be taken into consideration.

12.4   A 'debit' bank balance in the firm's cash book means that there is money at the bank.

12.5   **a**   Cheques not presented yet for payment – those cheques given or sent to creditors who have not yet claimed or collected payment from your bank.
   b   Cheques and cash not yet cleared – remittances (currency and cheques) paid into your own bank, but not yet shown on your bank statement. Often this is the total of the last paying-in slip handed in to the bank on the last day of the month.
   c   In the red – you are overdrawn.

13.1   The petty cash remaining is £13.57 so £16.43 is needed to top it up.

13.2   The petty cash remaining is £7.41 so £27.59 is needed to top it up.

13.3   The answer to the first part is in the first key point.

13.4   Using a separate petty cash book stops the main cash book getting cluttered up with lots of small entries.

13.5   A 'sundries' or 'miscellaneous' column takes care of casual and incidental expenses.

14.1 The purchases day book balance is £2035, creditors £1110.

14.2 The purchases day book balance is £1416, cash £1234, bank £5591, trial balance £14 624.

14.3 See 3.5.

14.4 The two largest categories of accounts are those of suppliers and of customers. These are therefore normally separated out into two separate ledgers – the purchases ledger and the sales ledger.

14.5 Subsidiary books called day books or journals are used for the purchases and sales of goods or services on credit. These books, referred to as original records of entry, are used as aids for posting the ledger accounts.

14.6 Trade discount is most often used by a supplier selling on to other traders but not to the public.

Cash discount, on the other hand, is a discount for prompt payment.

15.1 The purchases day book balance is £860, sales day book £933, cash £200 and bank £5581. The trial balance should total £11 322.

15.2 Sales day book is used to record all credit sales.

15.3 Cash sales are posted to the nominal ledger.

16.1 There should be a debit balance of £722.

16.2 Credit notes – see 2.1. Debit notes – see 2.1.

16.3 Returns outwards book – used to record purchase returns.

Returns inwards book – used to record sales returns.

17.1   VAT (Value Added Tax) is a tax on goods and services. It is intended to be a tax on the value added at each stage, until the product or service is bought by the consumer.

PAYE (Pay As You Earn) is the system under which the income tax and national insurance contributions due from employees are deducted from their pay by their employer who then pays it on to HM Revenue and Customs.

17.2   The level of turnover a business must register for VAT in 2009 is when its taxable supplies in the previous twelve months exceed £68 000.

17.3   In terms of VAT inputs are supplies that the business pays for and outputs are the supplies it makes.

17.4   A business should produce quarterly VAT returns, submitting them together with any payment on or before the end of the month following the end of the quarter. So, for example, a business on a January/April/July/October quarterly cycle would have to deliver a return for the three months to January by the end of February.

17.5   Gross pay is the amount before any tax is deducted, and net pay is the amount after both income tax and national insurance have been deducted.

18.1   First five are nominal, next four are real, last is personal.

18.2   £95 credit balance.

18.3   £409 debit balance.

Revision exercise 2

1   Cash £300, bank £19 901, trial balance should total £38 794.

2   Esther Allen is the debtor. 10th: £960 Cr, £600 Dr; 13th £2160 Cr, £600 Dr.

19.1 Gross profit is £2644, net profit £1964. The proprietor's closing capital is £31 064.

19.2 The nominal accounts form the trading and profit and loss account; the real accounts go to the balance sheet. The personal accounts are totalled to form debtors and creditors, and these go to the balance sheet as well.

19.3 Trial balance should total £11 024. The gross profit is £4939 and the net £2587. The net current assets (after liabilities) is £2607 and the proprietor's closing capital £3787.

20.1 Gross profit on sales 40%, net profit 30%, stock turn 5 times.

20.2 KC – the reduced mark-up is more than compensated for by the increased sales, shown by the faster stock turn. Remember that the gross profit percentage is NOT the same as the mark up – gross profit for RS is 40/140 × 100 = 28.57%.

20.3 5000 – (350 + 200) = 4450/2000 2.23 times (the operating profit can cover the interest charges 2.23 times).

Revision exercise 3 Cash £25, bank £457. Gross profit £57, net loss £74. Proprietor's capital £5826.

22.1 Gross profit £3294, net profit £1828

23.1 Capital £7500

23.2 2nd Dr Machinery, Cr Bank; 5th Dr Purchases Cr Cash; 12th Dr Fircone, Cr Sales; 15th Dr Purchases, Cr Heap; 20th Dr Cash, Cr Machinery; 22nd Dr Stationery, Cr Cash; 27th Dr Drawings, Cr Bank; 31st Dr Trading, Cr Stock.

23.3 1st Dr Machinery, Cr Purchases; 3rd Dr Sarah, Cr Sue; 6th Dr JA, Cr AJ; 12th Dr Drawings, Cr Salaries; 15th Dr Repairs,

Cr Machinery; 18th Dr Office equipment, Cr Stationery; 28th Cr Purchases £18; 30th Cr Bank interest £8.50, Cr Bank charges £8.50.

Revision exercise 4 Cash £200, Bank £4210, gross profit £2290, net profit £1680, proprietor's capital £57 680.

24.1 £180 $\dfrac{(3000 - £300)}{15}$

24.2 Lease £6400, depreciation £800 and £800. Machinery £2890, depreciation £600 and £510.

24.3 Net profit £3814, depreciation £1166, capital £7314.

25.1 Increase in provision £75, bad debts £8.

25.2 Provision decrease £70, bad debts £54.

25.3 Provision decrease £140, bad debts £300.

26.1 In advance £660, revenue £720.

26.2 Revenue £19 106, creditor for wages £662.

26.3 Revenue £218, creditors £38, revenue £195, in advance £65.

Examination exercise 2 Claim £3510. **3** Gross profit £7442, net profit £2903, capital £9923. **4 a** iii; **b** ii. **5** Asset balance £612. **6** 20 Main Street.

27.1 Net profit £3792.

27.2 Net profit £240.

27.3 Net profit £14 070.

28.1 Cash book balance b/f £172.10.

28.2 Surplus £411.

29.1 Dr £40, Cr £44 205.

29.2 Bought ledger £3207, Sales ledger £3939.

30.1 M £790, L £560.

30.2 A £1100, B £500, C £100.

30.3 SP £8700, MD £3700, BS £14 150.

Revision exercise 5

1 Cash book £7440, subscriptions £9310, deficit £1970.

2 Gross profit £8995, net profit £5065, net current assets £3630, current accounts JM £390, PR £(150).

32.1 Cost of sales £160 000: £140 000; gross profit 33.5%:30%; stock turn 6.4:5.5.

32.2 a £18 320; b £33 500; c £11 450; d £44 880.

32.3 a £8000; b £26 198; c 4252; d £32 105; e £14 447.

35.1 Published section retained profit c/f £11 540.

35.2 Net current assets £3610; total net assets £59 660.

Index